G000155851

EMERGING TECHNOLOGY FOR BIOREMEDIATION OF METALS

Edited by

Jeffrey L. Means and Robert E. Hinchee
Battelle, Columbus, Ohio

Library of Congress Cataloging-in-Publication Data

Catalog record is available from the Library of Congress

Lewis Publishers is an imprint of CRC Press

International Standard Book Number 1-56670-085-X
Printed in the United States of America 1 2 3 4 5 6 7 8 9 0
Printed on acid-free paper

CONTENTS

Articles

Technical Notes

FOREWORD

The use of bioremediation to treat metals-contaminated soil and water is an emerging field. The argument can be made that, for many years, conventional activated sludge treatment removed metals from wastewater. Although this is true, in most cases it was an inadvertent effect generally unrecognized and not monitored, and to the editors' knowledge, never optimized.

As an emerging discipline, metals bioremediation shows promise in a number of areas; but its ultimate utility is difficult to predict. Bioremediation traditionally has been applied to organic compounds that are broken down into harmless components, usually carbon dioxide and water. Applying this technology to metals presents a very different problem. Metals are elements that cannot be biologically degraded. Bioremediation of metals depends upon sorption, incorporation into more complex compounds, or valence state change. Sorption can be utilized to remove metals, resulting in substantial volume reduction. Incorporation into a more complex compound usually is a means of chemically stabilizing a metal. Metals removal by the activated sludge process works by sorption and, to some extent, incorporation into more complex compounds. Some metal organic compounds, such as methyl selenite, may be volatile, allowing removal by air stripping.

Bioremediation can lower the redox potential reducing some metals, which may lower solubility. For example, by reducing hexachrome to trichrome, chromium solubility and toxicity can be significantly reduced. The organisms do not act directly on the chromium but instead alter the environment in situ so that the metal is reduced to a less problematic state.

This book is intended to introduce the reader to the subject of metals bioremediation and is not an all-inclusive treatise. Given the early stage of development most metals bioremediation technologies currently are in, the state of the practice will evolve rapidly in the next few years, perhaps making such a treatise possible in 5 to 10 years.

This book is one of five volumes arising from the Second International Symposium on In Situ and On-Site Bioreclamation held in San Diego, California, in April 1993. The other volumes are *Bioremediation of Chlorinated and Polycyclic Aromatic Hydrocarbon Compounds*, *Hydrocarbon Bioremediation*, *Applied Biotechnology for Site Remediation*, and *Air Sparging*.

The symposium was attended by more than 1,100 people. More than 300 presentations were made, and all presenting authors were asked to submit manuscripts. Following a peer review process, 190 papers are being published. The editors believe that these volumes represent the most complete, up-to-date works describing both the state of the art and the practice of bioremediation.

The symposium was sponsored by Battelle Memorial Institute with support from a wide variety of other organizations. The cosponsors and supporters were:

Bruce Bauman, *American Petroleum Institute*
Christian Bocard, *Institut Français du Pétrole*
Rob Booth, *Environment Canada, Wastewater Technology Centre*
D. B. Chan, *U.S. Naval Civil Engineering Laboratory*
Soon H. Cho, *Ajou University, Korea*
Kate Devine, *Biotreatment News*
Volker Franzius, *Umweltbundesamt, Germany*
Giancarlo Gabetto, *Castalia, Italy*
O. Kanzaki, *Mitsubishi Corporation, Japan*
Dottie LaFerney, *Stevens Publishing Corporation*
Massimo Martinelli, *ENEA, Italy*
Mr. Minoru Nishimura, *The Japan Research Institute, Ltd.*
Chongrak Polprasert, *Asian Institute of Technology, Thailand*
Lewis Semprini, *Oregon State University*
John Skinner, *U.S. Environmental Protection Agency*
Esther Soczo, *National Institute of Public Health and Environmental Protection, The Netherlands*

In addition, numerous individuals assisted as session chairs, presented invited papers, and helped to ensure diverse representation and quality. Those individuals were:

Bruce Alleman, *Battelle Columbus*
Christian Bocard, *Institut Français du Pétrole*
Rob Booth, *Environment Canada, Wastewater Technology Center*
Fred Brockman, *Battelle Pacific Northwest Laboratories*
Tom Brouns, *Battelle Pacific Northwest Laboratories*
Soon H. Cho, *Ajou University, Korea*
M. Yavuz Corapcioglu, *Texas A&M University*
Jim Fredrickson, *Battelle Pacific Northwest Laboratories*
Giancarlo Gabetto, *Area Commerciale Castalia, Italy*
Terry Hazen, *Westinghouse Savannah River Laboratory*
Ron Hoeppel, *U.S. Naval Civil Engineering Laboratory*

Yacov Kanfi, *Israel Ministry of Agriculture*
Richard Lamar, *U.S. Department of Agriculture*
Andrea Leeson, *Battelle Columbus*
Carol Litchfield, *Keystone Environmental Resources, Inc.*
Perry McCarty, *Stanford University*
Blaine Metting, *Battelle Pacific Northwest Laboratories*
Ross Miller, *U.S. Air Force*
Minoru Nishimura, *Japan Research Institute*
Robert F. Olfenbuttel, *Battelle Columbus*
Say Kee Ong, *Polytechnic University, New York*
Augusto Porta, *Battelle Europe*
Roger Prince, *Exxon Research and Engineering Co.*
Parmely "Hap" Pritchard, *U.S. Environmental Protection Agency*
Jim Reisinger, *Integrated Science & Technology*
Greg Sayles, *U.S. Environmental Protection Agency*
Lewis Semprini, *Oregon State University*
Ron Sims, *Utah State University*
Marina Skumanich, *Battelle Seattle Research Center*
Jim Spain, *U.S. Air Force*
Herb Ward, *Rice University*
Peter Werner, *University of Karlsruhe, Germany*
John Wilson, *U.S. Environmental Protection Agency*
Jim Wolfram, *Montana State University*

The papers in this book have been through a peer review process, and the assistance of the peer reviewers is recognized. This typically thankless job is essential to technical publication. The following people peer-reviewed papers for this volume:

William Adams, *Monsanto Company U4E*
Edward R. Bates, *U.S. Environmental Protection Agency*
Hans Breukelman, *DSM Research BV, The Netherlands*
Jason A. Caplan, *ESE Biosciences*
Peter J. Chapman, *U.S. Environmental Protection Agency*
Soon H. Cho, *Ajou University, Korea*
Craig Criddle, *Michigan State University*
G. B. Davis, *CSIRO Division of Water Resources, Australia*
Wayne C. Downs, *U.S. Environmental Protection Agency*
J. A. Field, *Wageningen Agricultural University, The Netherlands*
Paul E. Flathman, *OHM Remediation Services Corporation*
Ian V. Fry, *Lawrence Berkeley Laboratory*
Paul Hadley, *California Environmental Protection Agency*
Duane D. Hicks, *Texas A&M University*

Brian S. Hooker, *Tri-State University*
Sydney S. Koegler, *Battelle Pacific Northwest Laboratories*
Raj Krishnamoorthy, *Aluminum Company of America*
Michael D. Lee, *DuPont Environmental*
Richard F. Lee, *Skidaway Institute of Oceanography*
Alfred P. Leuschner, *ReTec*
M. Tony Lieberman, *ESE Biosciences*
John Lyngkilde, *Technical University of Denmark*
Joan Macy, *University of California*
Pryodarshi Majumdar, *Tulane University*
Richard Ornstein, *Battelle Pacific Northwest Laboratories*
William C. Tacon, *Battelle Columbus*
J. van Eyk, *Delft Geotechnics, The Netherlands*
Richard Watts, *Washington State University*

The editors wish to recognize some of the key contributors who have put forth significant effort in assembling this book. Lynn Copley-Graves served as the text editor, reviewing every paper for readability and consistency. She also directed the layout of the book and production of the camera-ready copy. Loretta Bahn worked many long hours converting and processing files, and laying out the pages. Karl Nehring oversaw coordination of the book publication with the symposium, and worked with the publisher to make everything happen. Gina Melaragno coordinated manuscript receipts and communications with the authors and peer reviewers.

None of the sponsoring or cosponsoring organizations or peer reviewers conducted a final review of the book or any part of it, or in any way endorsed this book.

Jeff Means
Rob Hinchee
June 1993

BIOLOGICAL TREATMENT OPTIONS

L. A. Smith, B. C. Alleman, and L. Copley-Graves

ABSTRACT

Biological treatment technologies for metals remediation are in their infancy. Metals and their salts have the capacity to inhibit biological activity. Although some metals are essential micronutrients, they can become toxic to microorganisms at higher concentrations. Certain microorganisms have developed adsorption, oxidation, reduction, or methylation mechanisms to protect themselves from the toxic effects of metals. These mechanisms can be manipulated in treatment methods for metal contamination. Technologies that incorporate these mechanisms include biosorption, bioleaching and bioextraction, biobeneficiation, and biological oxidation or reduction. Although these technologies have been available for use in other areas, their potential for use in metals remediation has only recently gained attention.

INTRODUCTION

Microbial processes play a major role in the global cycling of metals. Microbial transformations are known to include both redox conversions of inorganic forms and transformations between inorganic and organic forms. Other microbial processes affect the distribution of metals by increasing or decreasing the solubility through various complexation reactions, by altering the pH of the environment, or through adsorption and/or uptake.

Metals and their salts have the capacity to inhibit biological activity. Although metals such as cobalt, copper, iron, manganese, molybdenum, and zinc are essential micronutrients, they can become toxic to microorganisms at higher concentrations, with the sensitivity being species dependent. Some microorganisms have developed adsorption, oxidation, reduction, or methylation mechanisms to protect themselves from the toxic effects of metals. These mechanisms can be manipulated in treatment methods for metal contamination.

Biological technologies are commonly used to treat certain types of organic contamination, but biological treatment technologies for metals remediation, with the exception of biosorbents, are in their infancy. Bioleaching and biobeneficiation have been used commercially in some mining operations, with biobeneficiation used to recover a variety of sulfidic ores. Adapting this type of biological treatment could prove useful in the recovery of metal values from soil and sludge wastes. These applications and others in the early stages of development are reviewed in this paper.

Biological activity can be exploited in several ways to remediate metals contamination. Biological processes can be used to alter the chemical state, form, or distribution of metals. Treatment can be conducted in situ or in more controlled scenarios such as aboveground reactor configurations. Currently, most of the potential techniques for biological treatment of metals contamination are being developed at the bench-scale level.

The more common biological mechanisms that can be exploited for remedial purposes include adsorption, uptake, oxidation or reduction reactions, and methylation/demethylation reactions. Technologies that incorporate these mechanisms include bioleaching and bioextraction, biobeneficiation, biosorption, and biological oxidation or reduction. These technologies have been available for use in other areas. However, their potential for use in metals remediation has only recently gained attention. The following sections describe these technologies.

BIOLOGICAL TREATMENT TECHNOLOGIES FOR METALS REMEDIATION

Bioaccumulation

Bioaccumulation involves the transfer of a metal from a contaminated matrix to biomass. Microbial biomass has been shown to adsorb inorganic as well as organic compounds from the aqueous phase. Metals can be accumulated by selected living organisms or onto inactivated nonliving biomass. The mechanisms to remove heavy metals from water can include two distinct pathways depending on whether the cells are living or dead. Biomass has been shown to be as efficient as many ion exchange resins for this type of removal. The process can be used to concentrate and recover metals, primarily from aqueous solutions.

Certain plant species and microorganisms can actively accumulate metals. Living cells also can adsorb metals this way and can concentrate inorganics within the cell. Although heavy metals may not be metabolically essential, they are taken up by the biomass as a side effect of the normal metabolic activity of the cells. Actively metabolizing biomass can concentrate metals by a variety of mechanisms including ion exchange at the cell walls, complexation reactions at the cell walls, intra- and extracellular precipitation, and intra- and extracellular complexation reactions. The net result of any of these processes is the sequestering of metals from solution (Benemann & Wilde 1991).

Inactivated biomass removes metals primarily through adsorbing metals to the ionic groups either on the cell surface or in the polysaccharide coating found on most forms of bacteria. The metals are bound by exchange of functional groups or by sorption on polymers. This process includes heavy metals which can be concentrated many times. The binding sites on the cell surface typically are carboxyl residues, phosphate residues, S-H groups, or hydroxyl groups.

Oxidation/Reduction

Selected microorganisms can oxidize or reduce metals. This type of activity can be used to remediate various types of metal contaminants. The oxidation or reduction of a metal may be performed directly by the organism, or may be a result of a reducing agent produced by the organism.

Microbially mediated redox reactions are defined as reactions through which an exchange of electrons results in a change in the valance of the metal. These types of reactions can be classified as either oxidation reactions that involve the removal of electrons from the metal and result in an increase in the valance state, or as reduction reactions in which electrons are removed from the metal atom and the valance state is reduced. The following examples show some common microbially mediated redox reactions.

Oxidation reactions:

mercury	Hg°	---->	Hg^{2+}	+	$2e^-$
cadmium	Cd	---->	Cd^{2+}	+	$2e^-$
arsenite	As^{3+}	---->	As^{5+}	+	$2e^-$
iron	Fe	---->	Fe^{2+}	+	$2e^-$
ferrous iron	Fe^{2+}	---->	Fe^{3+}	+	$2e^-$
manganese	Mn^{2+}	---->	Mn^{7+}	+	$2e^-$
antimony	Sb^{3+}	---->	Sb^{5+}	+	$2e^-$

Reduction reactions:

arsenate	As^{5+}	+	$2e^-$	---->	As^{3+}				
ferric iron	Fe^{3+}	+	$1e^-$	---->	Fe^{2+}				
mercury(I)	Hg^{1+}	+	$1e^-$	---->	Hg°				
mercury(II)	Hg^{2+}	+	$1e^-$	---->	Hg^{+}	+	$1e^-$	---->	Hg°

An increase in mobility is one result of microbial oxidation or reduction of metals that can be exploited for remedial purposes. For example, an insoluble sulfide may be oxidized to form a soluble sulfate. The solubilized form of the metal then dissolves from the matrix and can be collected for subsequent treatment, disposal, or recovery. Using this type of biological activity to effect an increase in mobility is analogous to soil washing or chemical leaching.

For some types of metal contaminants, biological oxidation or reduction may be used to reduce mobility or toxicity. For example, a limited number of bacteria can reduce hexavalent chromium, Cr(VI), to the less toxic trivalent form, Cr(III).

Indirect microbial transformation of metals can occur as a result of sulfate reduction when anaerobic bacteria oxidize simple carbon substrates with sulfate serving as the electron acceptor. The net result of the process is the production of hydrogen sulfide (H_2S) and alkalinity (HCO_3^-). Sulfate reduction is strictly an anaerobic process and proceeds only in the absence of oxygen. The process also requires a source of carbon to support microbial growth, a source of sulfate, and a population of sulfate-reducing bacteria. The H_2S formed in the reaction can react with many types of contaminant metals to precipitate metals as insoluble metal sulfides. The process follows the reaction

$$H_2S + M^{2+} \text{------>} MS(s) + 2H^+ \qquad (1)$$

The production of alkalinity from sulfate reduction reactions causes an increase in pH, which can result in metal removal through the formation of insoluble metal hydroxides or oxides. This process follows the reaction

$$M^{3+} + 3H_2O \text{------>} M(OH)_3(s) + 3H^+ \qquad (2)$$

In situ treatment of metals by stimulating sulfate reduction results in the precipitation of the metals and prevents further migration in contaminated soils or groundwater. When using sulfate reduction for metal removal from contaminated water, reactor systems must include a method for separating the resultant metal precipitates from the water stream.

Methylation

Biological methylation refers to the process in which organisms attach a methyl group ($-CH_3$) to an inorganic form of a metal. The biochemistry of methylation mechanisms is best understood from research investigating the methylation of mercury. Methylation results in the formation of organometallic compounds that are more volatile than the elemental form. The organometallic compounds can be removed from a contaminated matrix by volatilization and collected from the gas stream.

Methylation to remove arsenic from soils and sediments has been tested at the bench-scale level (Mattison 1993). However, it is unlikely that biological methylation will be employed at contaminated sites in the near future. Although methylation can increase the volatilization of a metallic contaminant, it results in a more toxic by-product that can be more difficult to control.

Methylation directly affects the mobility of the metal. Among metals known to be methylated, or demethylated, by microorganisms are mercury, arsenic, cadmium, and lead. The methylated forms of these metals typically are more mobile than the nonmethylated forms. The more mobile forms may leach from a site and impact surface or groundwater or may pose an air emission problem.

PERFORMANCE OF BIOLOGICAL TREATMENT

Applicability

Application of biological treatment of metals thus far has been limited to treatment of special wastes, because the characteristics of biological alternatives for metal treatment generally are unfavorable for broad application to waste treatment. Each biological treatment alternative for metals has a chemical analog. The biological treatment system typically operates at a slower rate than the analogous chemical mechanism. Biological systems that rely on direct action of live organisms are less tolerant of high metal concentrations. However, biological systems offer potential advantages over chemical treatment technologies.

Biological metal sorbents have similar applications to conventional water treatment processes such as ion exchange or activated carbon. In the case of inactive biosorbents, the material would appear to the end user much like conventional ion exchange resins. The advantage offered by biosorbents is higher loading capacity at low-metal-contaminant levels. Therefore, the biological materials effectively treat lower concentration influents and/or reach lower effluent concentrations. Biosorbents are also more selective for transition and heavy metals. Biomaterials may, therefore, be more effective for treatment of waters with high sodium or magnesium ion concentrations (Mattison 1993).

Biological oxidation/reduction treatment avoids the addition of the toxic chemicals that often are used in chemical oxidation/reduction or leaching processes. An otherwise chemically feasible in situ leaching or oxidation/reduction process may be unacceptable due to limitations on the types of chemicals that can be injected. Aboveground treatment allows more flexibility in the selection of chemicals but produces an aqueous wastestream that requires regeneration or cleanup and disposal. When available, biological mechanisms can achieve leaching, oxidation, or reduction without the addition of acids, bases, or oxidation/reduction agents.

Current Status of Technologies

Biological treatment of metals is generally less mature than most other metal treatment technologies. Biotreatment is, however, at the commercial or pilot scale of development for some metals.

Metal Biosorbent Materials. Both active and inactive biomass have been shown to have an affinity for metal ions. The commercial biosorbent materials are all based on inactive biomass. Inactive biomass materials that have been tested include algae (Jeffers et al. 1991) and a natural biopolymer, chitosan, made from shellfish wastes (Rorrer & Way 1991). Commercial biosorbents are summarized in Table 1.

Biosorption by living systems is being tested but has not reached commercial maturity. Nickel has been removed from metal plating wastes by *Enterobacter* and *Pseudomonas* species (Wong & Kwok 1992). Organisms are being genetically engineered to sequester metals such as cadmium, cobalt, copper, and mercury (Smit & Atwater 1991).

Bioreduction of Mercury Salts to Metal. Biological activity has been demonstrated to recover mercury by reducing the mercury salts to metal. Biological processes can reduce Hg(II) to mercury metal and in some cases may hydrolyze organomercuricals. Bioreactors are reported to be able to process feedwater containing 1 to 2 mg/kg of Hg(II) to give an effluent with 50 mg/kg or lower Hg(II). The mercury metal formed can be volatilized from the water stream by air stripping (Mattison 1993). Several organisms, including *Pseudomonas putida* (Horn et al. 1992) and *Thiobacillus ferrooxidans* (Hansen & Stevens 1992), have been tested for application to chemical reduction and recovery of mercury from wastewater. Bioreduction has been demonstrated at the bench-scale level but has not yet been taken to the pilot-scale level.

TABLE 1. Characteristics of commercial biosorbents.

Product	Developer	Biosorbent Type	Support Material	Metals Treated	Notes
MRA	AMT-BIOCLAIM	Caustic treated killed bacteria	Polymer	Available in cationic form and anionic form	
AlgaSORB	Bio-Recovery	Algae	Silica gel	Cationic metals	Tolerates low pH and high temperatures better than conventional ion exchange materials.
BIO-FIX	U.S. Bureau of Mines and licensees	Peat moss, *Spirulina*, and other	Polysulfone polymer	Cationic metals	

AMT = Advanced Mineral Technology
BIO-FIX = Biomass Foam Immobilized Extractant
MRA = Metal Removal Agent

Sources: Mattison, 1993; U.S. EPA, 1992, and U.S. EPA, 1990.

Bioreduction of Cr(VI) to Cr(III). As mentioned above, hexavalent chromium is a frequent candidate for chemical or electrochemical reduction to reduce its toxicity and mobility. Biological mechanisms can reduce Cr(VI) either by direct microbial action or by microbial production of sulfide.

Allied-Signal Research and Technology has developed a process using anaerobic sulfate-reducing bacteria to produce hydrogen sulfide (H_2S) which in turn reduces Cr(VI) to Cr(III). Formation of H_2S is a natural by-product of the metabolism of the sulfate-reducing bacteria. The pH-adjusted feed enters an upflow anaerobic sludge blanket column reactor. The bacteria coat the gravel in the distribution layer. As the feed passes up through the bacteria in the sludge layer, Cr(VI) is reduced to Cr(III) by the H_2S. Because the bacteria do not metabolize the chromium directly, they are somewhat protected from its toxic effects.

Several organizations are studying bioreduction of chromium by direct metabolism. In these processes, the Cr(VI) diffuses to the cell wall, binds with an enzyme, and is reduced. One variety of pure strain bacteria and bacterial consortia has been found to have the capability to reduce chromium. At the current stage of development, direct reduction gives generally slower reaction rates and higher effluent Cr(VI) residual than does indirect reduction.

The University of Tokyo is developing a dialysis culture system to help protect chromium-reducing bacteria from the toxic effects of chromium. The feed is separated from the anaerobic bioreactor by an anion-selective membrane. Chromate diffuses through the membrane and is reduced by the bacteria to Cr(III), which exists in solution as a cation and cannot diffuse back across the membrane.

Metal Treatment in Wetlands. Metal remediation in wetlands involves constructing and/or managing a wetland environment to encourage concentration of metal contaminants. The wetlands-based treatment technique combines natural geochemical and biological processes to accumulate and remove metals from influent waters. The treatment system incorporates principal ecosystem components of the wetland, including organic soils, microbial fauna, algae, and vascular plants to reduce the aqueous concentrations of metals. Most of the metals remediation is due to microbial activity.

The wetland typically has an aerobic zone at the surface and an anaerobic zone below the surface. Oxidation and reduction reactions catalyzed by bacteria play a major role in metals removal in both zones. Under aerobic conditions, metals with insoluble oxides are removed. Anaerobic processes, such as sulfate reduction, remove metals that form insoluble sulfides.

Wetlands-based treatment was studied under the U.S. Environmental Protection Agency's Superfund Innovative Technology Evaluation (SITE) emerging technology program. Demonstration of a full-scale constructed wetland is planned under the SITE demonstration program. The planned test site is the Burleigh Tunnel, which is part of the Clear Creek/Central City Superfund site in Colorado (U.S. EPA, 1992).

Bioleaching. Bioleaching involves microbiological solubilization of metal contaminants to improve removal from a solid or semi-solid matrix. Bioleaching may be used to recover metals from either in situ or excavated material. The organisms dissolve materials by direct action of active bacteria, by indirect attack by one or more metabolic products, or by a combination of both direct and indirect attack.

Bioleaching processes make use of current mining practice. Bioleaching is now applied for heap and in situ leaching. New bioleaching techniques will allow heap leaching to recover different types of metals. New bioreactors are being developed to compete with chemical reactors to leach metals from ores.

Bioleaching to recover metal values on an industrial scale is currently practiced for extracting copper and uranium (Ehrlich 1988) by heap or in situ leaching of raw ore. The ore typically is low in metal content, e.g., less than 0.5% copper. Bioleaching of ore concentrates has attracted some interest but would best be accomplished in a bioreactor. Bioreactors for treating higher grade copper ores are under development at the pilot scale.

The copper and uranium ores treated by bioleaching are both sulfide rich (Brierly 1990). Leaching is believed to be accomplished by a complex combination of *Thiobacillus ferrooxidans*, *Leptospirillum ferrooxidans*, *Thiobacillus organosparus*, and probably other Gram-negative acidophilic bacteria (Hinsenveld 1991). The exact system of organisms and mechanisms has not yet been fully defined.

Bioleaching is under development at the bench or pilot scale for a wide range of metals. The most advanced developments are in heap leaching of gold and silver pyrites or arsenopyrite ores (Attia & Elzeky 1991, Mining Engineering 1984). Other metals under study for bioleaching include aluminum, cadmium, chromium, iron, lead, manganese, mercury, molybdenum, nickel, selenium, and tin (Brooks et al. 1991, Shall 1992).

Recovering mercury by microbial activity has recently received attention. Mercury salts can be reduced to elemental metal through biological activity. The native mercury metal is a dense liquid and is easily separated for aqueous streams after bioreduction. *Pseudomonas putida* (Horn et al. 1992) and *Thiobacillus ferrooxidans* (Hansen & Stevens 1992) have been tested for application to chemical reduction and recovery of mercury from wastewater.

Biobeneficiation. In biobeneficiation, microorganisms improve the physical separation of a contaminated solid matrix into contaminant-rich and contaminant-poor streams. Beneficiation produces mineral concentrate and tailings streams without substantial chemical changes to the matrix being processed. The physical beneficiation occurs after the excavated waste is processed to reduce particle size. The separation can be based on a variety of physical or chemical properties such as color, luster, particle size or shape, specific gravity, magnetic permeability, inductive charging, or surface chemical properties.

Biobeneficiation typically is applied to improve the performance of separation by froth flotation. Froth flotation depends on a difference in affinity for water between the surface of the ore and the tailings particles. Air is bubbled through a slurry of crushed rock called the pulp. The bubbling action generates a froth on the surface of the pulp. Particles with a hydrophobic surface are attracted to the air bubbles and concentrate in the froth. Particles with a hydrophilic surface settle to the bottom. Chemical treatment typically is needed to increase or decrease the surface activity of various ore and tailings materials to maximize separation.

Biological action could be used to replace the chemical treatment required in conventional flotation processes. For example, in standard practice, cyanide ions are added to ore pulp containing iron sulfide (pyrite) to form an iron-cyanide complex. The presence of the complex on the surface prevents the pyrite particles from attaching to the bubbles. Biotreatment with *Thiobacillus ferrooxidans* has been studied as a method to modify ore surfaces to improve flotation processes to separate complex sulfide ores (Rao et al. 1992) and remove sulfide contaminants from coal (Attia & Elzeky 1985).

Biological action also can be applied to improve settling or filtration by controlling the agglomeration of mineral particles. The action of *Aspergillus niger* in suspensions of calcite, magnetite, and barite was studied to identify bioeffects on agglomeration behavior (Rao et al. 1992). *Mycobacterium pheli* has been shown to be a good flocculating agent for coal, phosphate slimes, and hematite, and *Mycobacterium pheli* also has been tested for recovery of kerogen (asphalt-like material) from oil shale (Shall 1992).

Most biobeneficiation techniques are specific to improving performance of physical separation in flotation and settling processes with sulfide ores.

Vegetative Uptake. Certain species of plants can concentrate metals by uptaking them through their root systems and depositing the metals in the leaves. For example, many plants in the Leguminosae family have specific mineral requirements and thrive only in habitats with those minerals. Plants in the *Astragalus* genus (locoweeds) often grow on shale soils containing selenium and barium instead of the usual sulfur. In dry years the roots penetrate deeper and take up more selenium in their search for water. Alfalfa and clover both require boron. Alfalfa and most beans require manganese, and alfalfa takes in potassium.

Choosing Biological Treatment

The applicability of any form of biotreatment and the performance of the process depend on the following conditions:

- type(s) and concentration(s) of metal(s)
- matrix
- pH
- temperature
- oxygen concentration
- alkalinity
- substrate availability
- nutrient concentrations
- presence of indigenous microorganisms
- population density
- cell age
- use of active or inactive biomass
- contact time.

Other properties affecting application of biotreatment of metals are both technology specific and site specific and vary on a case-by-case basis. These include the volume of contaminated material, depth of the contaminated material, and controllability of the site.

REFERENCES

Attia, Y. A., and M. Elzeky. 1991. "Effect of Bacterial Adaptation and Solution Replacement on Bioleaching of Sulfidic Gold Ores." *Minerals and Metallurgical Processing.* 9(3):122-127. August.

Attia, Y. A., and M. Elzeky. 1985. "Biosurface Modification in the Separation of Pyrite from Coal by Froth Flotation." In Y. A. Attia (Ed.), *Processing and Utilization of High Sulfur Coals*, pp. 673-682. Elsevier Science Publishing Company, Amsterdam.

Benemann, J. R., and E. W. Wilde. 1991. *Literature Review on the Use of Bioaccumulation for Heavy Metal Removal and Recovery*. WSRC-TR-90-175-Vol. 2. Westinghouse Savannah River Co., Aiken, SC. February.

Brierly, J. A. 1990. "Biotechnology for the Extractive Metals Industries." *Journal of Metals 42*(1):28-30. January.

Brooks, C. S., P. L. Brooks, G. Hansen, and L. A. McCarthy. 1991. *Metal Recovery from Industrial Waste*. Lewis Publishers, Chelsea, MI. pp. 113-117.

Ehrlich, H. L. 1988. "Recent Advances in Microbial Leaching of Ores." *Minerals and Metallurgical Processing 5*(2):57-60. May.

Hansen, C. L., and D. K. Stevens. 1992. "Biological and Physio-chemical Remediation of Mercury-Contaminated Hazardous Waste." *Arsenic and Mercury: Workshop on Removal, Recovery, Treatment, and Disposal*, pp. 121-125. EPA/600/R-92/105.

Hinsenveld, Ir. M. 1991. "Recent Developments in Extraction and Flotation Techniques for Contaminated Soils and Sediments." Presented at the Fifth International NATO/CCMS Conference on Demonstration of Remedial Action Technologies and Groundwater, Washington, DC. November 18-22.

Horn, J. M., M. Brunke, W. D. Deckwer, and K. N. Timmis. 1992. "Development of Bacterial Strains for the Remediation of Mercurial Wastes." *Arsenic and Mercury: Workshop on Removal, Recovery, Treatment, and Disposal*, pp. 106-109. EPA/600/R-92/105.

Jeffers, T. H., C. R. Ferguson, and P. G. Bennett. 1991. *Biosorption of Metal Contaminants Using Immobilized Biomass: A Laboratory Study*. BUMINES-RI-9340. Bureau of Mines, Washington, DC.

Mattison, P. L. 1993. *Bioremediation of Metals — Putting It To Work*. Cognis. Santa Rosa, CA.

Mining Engineering. 1984. "Bioleaching Method for Sulfidic Gold and Silver." *Mining Engineering 36*(12):1626. December.

Rao, M., K. Yelloji, K. A. Natarajan, and P. Somasundaran. 1992. "Effect of Biotreatment with *Thiobacillus ferrooxidans* on the Floatability of Sphalerite and Galena." *Minerals and Metallurgical Processing 9*(2):95-100. May.

Rorrer, G. L., and J. D. Way. 1991. *Chitosan Beads to Remove Heavy Metals from Wastewater*. U.S. Department of Energy, Innovative Concepts Program. Compiled by Raymond Watts, Pacific Northwest Laboratory, Richland, WA.

Shall, H. El. 1992. "Biotechnology." *Mining Engineering. 44*(5):460-462. May.

Smit, J., and J. Atwater. 1991. *Use of Caulobacters to Separate Toxic Heavy Metals from Wastewater Streams*. U.S. Department of Energy, Innovative Concepts Program. R. Watts (comp.), Pacific Northwest Laboratories, Richland, WA.

U.S. Environmental Protection Agency. 1990. *Technology Evaluation Report: SITE Program Demonstration Test, Solidtech, Inc. Solidification/Stabilization Process*, Vol. 1. EPA/540/5-89/005a. U.S. EPA SITE Program report. 120 pp.

U.S. Environmental Protection Agency. 1992. *The Superfund Innovative Technology Evaluation Program: Technology Profiles, 5th ed.* EPA/540/R-92/077. U.S. EPA Office of Solid Waste and Emergency Response. 388 pp.

Wong, P. K., and S. C. Kwok. 1992. "Accumulation of Nickel Ion by Immobilized Cells of Enterobacter Species." *Biotechnology Letters 14*(7):629-634.

PASSIVE BIOREMEDIATION OF METALS FROM WATER USING REACTORS OR CONSTRUCTED WETLANDS

T. R. Wildeman, D. M. Updegraff, J. S. Reynolds, and J. L. Bolis

ABSTRACT

Guidelines for microbial treatment of metal contaminants come from principles developed in aquatic geomicrobiology. Most treatment is controlled by microbial processes, so laboratory- and bench-scale testing, typical of other process design, can be utilized. One laboratory study confirmed that incubation for 4 weeks produced treatment results comparable to those from demonstration wetlands. The same study found that the rate of sulfide production in an anaerobic system started at 1.2 and decreased to 0.75 micromole of $S^{=}$/gm of substrate/day. Bench-scale studies were used to determine the loading capacity of the bioreactor, soil permeability for anaerobic reactors, and design suitability. For anaerobic reactors, loading is based on the concept that there should always be excess sulfide compared to heavy metals. In one anaerobic study that used a severely contaminated drainage, pH was increased from 2.5 to 7.5. Metal concentration changes in mg/L were as follows: Zn (150 to 0.2), Cu (55 to <0.05), Fe (700 to 1), and Mn (80 to 1). The concepts have been tested on anaerobic, pilot-scale passive reactors. The pH was increased from 3.0 to 6.5 and metal concentrations decreased to a lesser degree.

MICROBIAL GUIDELINES FOR METALS REMEDIATION

With an expenditure of energy, all metals have been extracted from ores found at the surface or near surface of the earth. Unlike organic compounds, metals cannot be destroyed. Because minerals form over geologic time, thermodynamics dictates that they represent the most stable chemical form for that metal. Consequently, the objective of metals treatment is to return contaminants to their natural mineral forms. Many of these minerals, such as FeS_2 and MnO_2, are formed from water in a sedimentary environment. Many of these chemical reactions that occur in an aquatic environment are catalyzed by bacteria. The bioremediation of metal contaminants involves optimizing what has been naturally occurring throughout geologic time.

When remediating metal contaminants in water, the objectives usually require adjusting the pH to about 7 and removing the metals as sulfide, hydroxide, or carbonate precipitates. These objectives generate one of the primary guidelines for metals bioremediation. Because $S^{=}$, $CO_3^{=}$, and OH^- are common products of bacterial activity, enzymatic uptake of metals into bacteria is not necessary for remediation. Consequently, a more general approach to microbiology is used in passive treatment of metal contamination.

Microbial processes in aerobic environments are very different from those in anaerobic environments. Aerobic conditions are effective in removing metals whose oxides are relatively insoluble. These include Fe(III), and Mn(IV). Anaerobic processes, including sulfate reduction, are effective in removing metals that form insoluble sulfides. These include Cu, Zn, Cd, Pb, Ag, and Fe(II). Both aerobic and anaerobic processes can neutralize acids, increasing the pH, and add alkalinity to water in the form of HCO_3^-. Consequently, in either environment, it is possible to remove Al and Cr(III) as hydroxides, or Zn and Cu as carbonates.

The most important aerobic biological processes in wetlands are iron oxidation and photosynthesis. Both are autotrophic processes in which carbon dioxide is the source of carbon for the organisms concerned. Photosynthesis, carried out by blue-green bacteria, algae, and plants, consumes carbonic acid and bicarbonate and produces hydroxyl ions:

$$6\ HCO_3^-\ (aq) + 6\ H_2O \rightarrow C_6H_{12}O_6 + 6\ O_2 + 6\ OH^-$$

In this case aquatic organisms are making organic matter by taking up dissolved bicarbonate to produce dissolved oxygen and hydroxide ions (Wetzel 1983).

The oxidation of iron pyrite by aerobic autotrophic bacteria of the genus *Thiobacillus* is the cause of acid mine drainage, as summarized by the following reactions from Stumm and Morgan (1981):

$$FeS_2\,(s) + 7/2\,O_2 + H_2O \rightarrow Fe^{2+} + 2\,SO_4^{=} + 2\,H^+$$
$$Fe^{2+} + \tfrac{1}{4}\,O_2 + H^+ \rightarrow Fe^{3+} + \tfrac{1}{2}\,H_2O$$
$$Fe^{3+} + 3\,H_2O \rightarrow Fe(OH)_3 + 3\,H^+$$
$$FeS_2 + 14\,Fe^{3+} + 8\,H_2O \rightarrow 15\,Fe^{2+} + 2\,SO_4^{=} + 16\,H^+$$

Note that H^+ is produced by the oxidation of bisulfide and by the precipitation of $Fe(OH)_3$. Manganese oxidation and precipitation also releases H^+:

$$2\,H_2O + Mn^{2+} \rightarrow MnO_2 + 3\,H^+ + 2\,e^-$$

Finally, oxidation of organic matter produces H^+:

$$H_2O + \text{"}CH_2O\text{"} \rightarrow CO_2 + 4\,H^+ + 4\,e^-$$

Here, "CH_2O" represents organic matter such as cellulose and other carbohydrates.

Under anaerobic conditions in wetlands, five general types of microbial processes are of importance:

1. Hydrolysis of biopolymers by extracellular bacterial enzymes. An example is the hydrolysis of cellulose, the most abundant organic material in plants, to glucose:

$$(C_6H_{11}O_5)_n + n\,H_2O \rightarrow n\,C_6H_{12}O_6$$

2. Fermentation; examples are the formation of ethanol and pyruvic acid:

$$C_6H_{12}O_6 \rightarrow 2\,C_2H_5OH + 2\,CO_2$$
$$C_6H_{12}O_6 \rightarrow 2\,C_3H_4O_3 + 4\,H^+$$

3. Methanogenesis:

$$CO_2 + 4\,H_2 \rightarrow CH_4 + 2\,H_2O$$

4. Sulfate reduction:

$$2\,H^+ + SO_4^{=} + 2\,\text{"}CH_2O\text{"} \rightarrow H_2S + 2\,H_2CO_3$$

5. Iron reduction:

$$Fe^{3+} + e^- \rightarrow Fe^{2+}$$

Fermentation often produces acids, decreasing pH, while sulfate reduction consumes H+ and increases pH. Methanogenesis consumes hydrogen ions also. Proton-reducing bacteria, which are symbionts with methanogenic bacteria, convert H+ to H_2, and this is used by the methanogenic bacteria to reduce CO_2 to CH_4. In our anaerobic wetland environments, sulfate reduction and methanogenesis proceed together. Because Postgate (1979) reports that the activity of sulfate-reducing bacteria is severely limited below pH 5, organic materials in a constructed wetland environment have to be chosen so that fermentation does not dominate over sulfate reduction.

MICROBIOLOGICAL GUIDELINES APPLIED TO CONSTRUCTED WETLANDS

Although still not completely understood, the principles outlined above appear to be the predominant removal mechanisms in the treatment of mine drainage and other metal-contaminated waters by constructed wetlands (Hammer 1989, Wildeman et al. 1992). In the early 1980s, aerobic removal processes were emphasized and the precipitation of $Fe(OH)_3$ was an important objective. Because precipitation of $Fe(OH)_3$ produces H^+ ions, iron was removed but the pH of the effluent often was around 3. Brodie (1991) has had success with metals treatment using aerobic constructed wetlands as long as the pH of the influent was above 5.5 and carried some alkalinity in the form of dissolved bicarbonate. Around 1987, groups from the U.S. Bureau of Mines (Hedin et al. 1989) and the Colorado School of Mines (Wildeman & Laudon 1989, Wildeman et al. 1992) began to investigate the role of anaerobic processes, particularly sulfate reduction, in treating acid mine drainage.

The microbial guidelines presented above have been the result of these early studies. Application of these guidelines to aerobic wetlands leads to the following four practices for success:

1. Aerobic removal processes are successful when the pH of the effluent water is above 5.5 and dissolved bicarbonate is present.
2. Any processes, such as anoxic limestone drains (Brodie et al. 1991), that will raise the pH and add alkalinity should be used.
3. Precipitation of iron and manganese oxyhydroxides is a primary removal process and other metal contaminants are removed by adsorption onto these precipitates on by precipitating as carbonates.

4. Plants are essential to success because photosynthesis is a primary process for raising pH, adding oxygen to the water, and supplying organic nutrients.

The role of photosynthesis can be best understood by considering the reaction given in the previous section. At about pH 5.5, significant amounts of bicarbonate can be retained in water (Stumm & Morgan 1981), and this appears to be why such wetlands are effective when the pH is above this value.

Application of the guidelines to anaerobic removal processes is much more direct, because plants are absent and the system is dominated by microbial processes. The following four practices lead to success:

1. Wetland substrates are formulated so that organic material necessary for metabolism is in high abundance and the soil can provide acid buffering capacity at a pH above 7.
2. Microbial processes that transform strong acids such as H_2SO_4 into weak acids such as H_2S are promoted.
3. The products of these reactions are used to precipitate metal contaminants as sulfides (CuS, ZnS, PbS, CdS), hydroxides ($Al(OH)_3$, $Cr(OH)_3$), and carbonates ($MnCO_3$).
4. To remain effective, the reactions that consume H^+ have to predominate over the reactions that produce H^+.

Using these guidelines results in the construction of systems that work more like passive bioreactors than wetlands (Wildeman 1992). However, they have been successful at raising the pH of metal mine drainages from below 3 to above 6 and have reduced metals concentrations (in mg/L) the following amounts: Fe, 30 to <1; Cu, 1 to < 0.03; and Zn, 9 to < 0.03. How anaerobic remediation systems are designed will be the focus of the rest of this paper.

STAGED DESIGN OF ANAEROBIC REACTOR SYSTEMS

In our studies on treatment of metal mine drainage by constructed wetlands (Wildeman et al. 1992), when it was determined that precipitation of metals by sulfide generated from sulfate-reducing bacteria is the important process, it was realized that establishing and maintaining the proper environment in the substrate is the key to success for removal.

This means that processes operating on the surface of the wetland may be neglected for design of anaerobic treatment systems. If this is the case, then construction of large pilot cells is not necessary to determine if a wetland that emphasizes anaerobic processes for removal will work.

Consequently, study of wetland processes and design of optimum systems can proceed from laboratory experiments, to bench-scale studies, and then to the design and construction of actual cells. We call this "staged design of wetland systems." Although staged design is best carried out on anaerobic substrates, it has also been used with success on design of aerobic systems. Algal photosynthesizers are excellent generators of oxygen and alkalinity in water, and they can be readily used in laboratory and bench-scale studies of aerobic treatment (Duggan et al. 1992). In actual wetland systems, the growth of glue-green bacteria, algae, and plants on the surface of the system may be important because they increase pH and produce organic matter essential for the growth of sulfate reducers and other heterotrophic bacteria in the subsurface.

Laboratory Studies. In early laboratory studies, culture bottle experiments were used for studies on how to establish tests to determine the production of sulfide by bacteria, and of what substrate will provide the best initial conditions for growth of sulfate-reducing bacteria (Reynolds et al. 1991). In particular, great emphasis is placed on testing local organic and soil materials to determine what mix provides the best environment for sustained sulfate-reducing bacterial activity. Recently, laboratory studies have concentrated on the practical aspects of wetland design. One industrial concern was interested in whether cyanide concentrations typical of milling-waste effluents would kill sulfate-reducing bacteria (Filas & Wildeman 1992). Culture bottle tests showed that sulfate reduction was retarded until the concentration of total cyanide was below 10 mg/L. Another industrial concern wanted to determine whether Cu concentrations above 100 mg/L would kill or retard sulfate-reducing bacteria. Culture bottle tests conducted over the course of 1 month showed that sulfate reduction was still vigorous at Cu concentrations above 100 mg/L.

In the most extensive laboratory study run, a series of culture bottles was sealed and incubated at 18°C to determine the activity of sulfate-reducing bacteria and whether the metals removal in the laboratory was comparable to that in a demonstration anaerobic reactor (Reynolds et al. 1991). For the laboratory study, 20 g of substrate and 70 mL of mine drainage, whose chemistry is shown in Table 1, were sealed into 120-mL serum bottles. The substrate came from an active anaerobic cell from the Big Five Pilot Wetlands in Idaho Springs, Colorado (Wildeman et al. 1992). The rate of sulfate reduction was measured at intervals by two

TABLE 1. Comparison of pH, sulfate, and metal concentrations in mine drainage, wetland output, and serum bottles adapted from Reynolds et al. 1991.

Sample	pH	$SO_4^=$ (mg/L)	Cu (mg/L)	Fe (mg/L)	Mn (mg/L)	Zn (mg/L)
Mine Drainage	3.0	1720	0.57	39	31	8.6
Reactor Effluent	6.7	1460	<0.05	0.64	15.8	0.07
Serum Bottles						
1 day (baseline)[a]	6.1	1680	<0.05	10.5	15.5	0.04
5 days[a]	6.2	1660	<0.05	7.3	10.8	0.53
10 days[b]	6.3	1610	<0.05	4.7	9.8	0.27
15 days[b]	6.3	1530	<0.05	8.8	15.3	0.13
20 days[b]	6.4	1410	<0.05	8.1	12.0	0.37
25 days[b]	6.4	1470	<0.05	6.9	11.3	0.22
30 days[b]	6.4	1350	<0.05	4.9	9.7	0.21
35 days[b]	6.7	1240	<0.05	4.51	10.6	0.16

(a) Values for these samples are the average of 4 replicates.
(b) Values for these samples are the average of 3 replicates.

different methods, one using $^{34}SO_4^=$ and one not requiring a radioactive tracer. On the day of collection, the effluent from the reactor had the chemistry shown in Table 1. The results for the metals removal are shown in Table 1.

In the baseline bottles, measured 1 day after adding the mine drainage, the pH was significantly higher and metals concentrations were significantly lower than in the mine drainage. As stated in the wetlands guidelines, substrates are formulated to immediately raise the mine drainage pH to neutral conditions. Several processes could have contributed to the immediate removal of metals, including precipitation due to the increase in pH and adsorption onto organic materials. Over the course of 35 days, the pH continues to rise until it is the same as in the effluent from the anaerobic reactor. The concentration of sulfate gradually decreases and sulfide increases until, by day 35, sulfate is below that in the reactor effluent. At 25 days the concentration of sulfate matches that in the reactor effluent. Based on this observation, laboratory-scale tests are conducted for at least 4 weeks to simulate conditions in the field. After 35 days, the pH of the laboratory solutions was the same as for the effluent, and metals removal was comparable. The laboratory studies mimic what is occurring in the demonstration reactor.

Because the serum bottles were sealed, all volatile products were retained. In particular, all retained sulfide species could be titrated and the amount of production gives an estimate of the activity of sulfate reduction. This rate for the first 40 days was 1.2 µM of sulfide/g of substrate/day (Reynolds et al. 1991). From 40 to 80 days, the rate averaged 0.75 µM/g/d. In design calculations for bench- and pilot-scale reactors, a sulfide production rate of 300 nM/g/d is used.

Bench-Scale Reactors. For bench-scale studies, plastic garbage cans are used to conduct experiments to provide answers necessary to the design of a subsurface cell (Bolis et al. 1991). Typical design parameters to be determined include the optimum loading factor, substrate, cell configuration, and the permeability of the substrate. In a bench-scale study recently completed, garbage cans filled with substrate were used to determine whether using the sulfide generation figure of 300 nM sulfide/cm^3 substrate/day could be used to set the conditions for treating severely contaminated drainage that flows from the Quartz Hill Tunnel in Central City, Colorado. Contaminant concentrations for this drainage are shown in Table 2. Using the practice guideline described above that the amount of sulfide produced should always be in excess of the amount of heavy metals to be precipitated, with the amount of substrate contained in the garbage can, flow could not exceed 1 mL/m to ensure that produced sulfide would always be in excess.

Contaminant concentrations from the outputs of three different bench-scale cells are shown in Table 2. For Cell A, the mine drainage was passed through the cell with no delay. For Cell B, the substrate was soaked with city water for 1 week before mine drainage started passing through the cell. For Cell C, the substrate was inoculated with an active culture of sulfate-reducing bacteria and soaked with city water for 1 week before mine drainage started passing through the cell. Preparations on Cells B and C were done to ensure that there would be a healthy population of sulfate-reducing bacteria before mine drainage flowed through the substrate. All cells were run in a downflow mode of the mine drainage flow through the substrate. In all three cells removal of Cu, Zn, Fe, and even Mn is greater than 99%. The pH increases from about 2.5 to above 7. These results were maintained consistently for more than 10 weeks of operation.

The substrate used was a mix of ¾ composted cow manure and ¼ planting soil. The results from Cells A and B show that the cow manure has an indigenous population of sulfate-reducing bacteria that are capable of utilizing the organic material in the manure. Inoculation with an active culture of bacteria is not necessary in this case. Also, because the results

TABLE 2. Constituent concentrations in mg/L in the Quartz Hill Tunnel mine drainage and in effluents from the bench-scale tests.

Sample	Days Operated	Mn	Fe	Cu	Zn	SO_4	pH
		Concentration in mg/L					
Mine Drainage	24	80.0	630.0	48.0	133.0	4240	2.4
Cell A	24	0.94	1.6	0.06	0.27	450	7.4
Cell B	24	0.91	1.9	<0.05	0.17	770	7.5
Cell C	24	0.99	1.0	<0.05	0.16	412	7.4
Mine Drainage	43	80.0	640.0	50.0	135.0	4300	2.5
Cell A	43	0.97	0.87	<0.05	0.18	1080	7.2
Cell B	43	0.64	0.96	<0.05	0.24	660	7.4
Cell C	43	1.6	0.46	<0.05	0.14	1180	7.2
Mine Drainage	71	70.0	820.0	70.0	101.0	NA	2.6
Cell A	71	0.48	0.40	<0.05	0.21	NA	8.0
Cell B	71	1.6	0.40	<0.05	0.25	NA	7.9

from Cell A are comparable to those of Cells B and C, the population of sulfate reducers can withstand immediate exposure to severe mine drainage and still produce sufficient quantities of sulfide. The key to good initial activity is to ensure that the flow of mine drainage is low enough that its low pH does not disturb the microenvironment established by the bacteria.

Another feature of the results shown in Table 2 is that Mn is removed in all three cells. Typically, Mn is the most difficult contaminant in mine drainage to remove (Brodie 1991, Duggan et al. 1992, Hedin et al. 1989, Wildeman et al. 1992). It is usually presumed that removal of Mn has to be achieved by raising the pH to above 7, and then introducing the effluent into an aerobic wetland cell so that Mn will be oxidized to MnO_2. Removal in an anaerobic cell must be as Mn(II). Analyses of possible species at a pH above 7 suggest that removal could be as MnS or $MnCO_3$.

Mn does not adsorb onto the organic material as readily as Fe and Cu (Machemer & Wildeman 1992). In this case, it is hypothesized that $MnCO_3$ is the precipitate because it is more insoluble than the sulfide. In either case, a key to Mn removal in an anaerobic cell appears to be the ability to raise the pH of the effluent above 7. If raising the pH to above 7 can be consistently achieved, then all the contaminants in mine drainage can be removed in one anaerobic cell.

Pilot-Scale Systems. Successful completion of bench-scale tests support a decision to proceed with pilot-scale testing. Bench-scale tests provide relevant results under field climatic and hydrologic conditions. If possible, the pilot cell is a module large enough that expansion to full-scale treatment involves adding more reactor cells. The objective is to collect data on the performance of a reactor module under actual site conditions.

The constructed wetland at the Big Five Tunnel in Idaho Springs, Colorado, is a good example of a pilot-scale facility that has been operated and performance monitored for more than 2 years (Wildeman et al. 1992). This system is an anaerobic wetland for treating acid mine drainage of the chemistry shown in Table 1. From 1987 to 1991, a number of cells have been built and tested. However the cell that was designed using the guidelines presented in this paper was Cell E.

Removal efficiencies for Cell E over a 27-month period are shown in Figure 1. Removal is determined by dividing the wetland effluent by the mine drainage influent concentrations. If removal is complete, the ratio will be close to zero. Examination of Figure 1 shows consistent and complete removal of Cu and Zn. Cell E was started in September 1989, so months 4 through 8 and 16 through 20 would be the winter. Removal of Fe changes with the seasons; it is good in the summer and poor in the

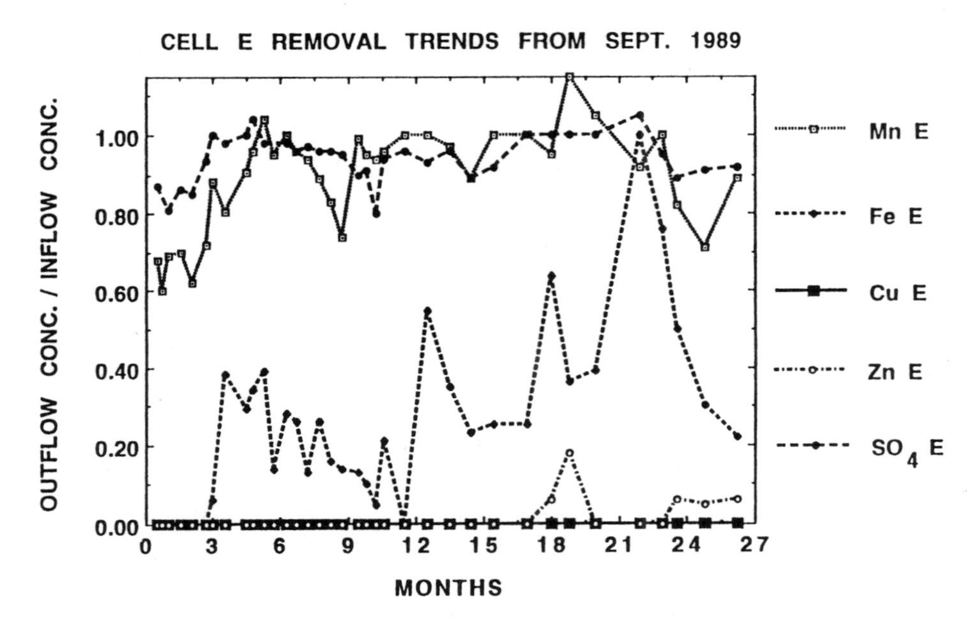

FIGURE 1. Cell E pilot-scale test results. Vertical axis is effluent over mine drainage concentration. The cell was started in September 1989.

winter. Because FeS is more soluble than CuS or ZnS, this winter increase is probably due to the reduction in activity of sulfate-reducing bacteria with decrease in temperature (Postgate 1979). Reduction of bacterial activity appears to be confirmed by the fact that there is a smaller decrease in sulfate concentration in the winter. Because Mn is the most soluble sulfide (Stumm & Morgan 1981), its removal is inconsistent. Note that the results from the pilot cell confirm the laboratory study results given in Table 1.

With the pilot-scale results, actual on-site removal can be assessed and the design criteria for full-scale treatment determined. With this record of treatment, a reasonable case can be presented on how well constructed wetlands and passive treatment reactors will operate in full-scale treatment of contaminated waters.

SUMMARY

Bioremediation of metals can be carried through a staged design process that is comparable to what is used in the bioremediation of fossil fuels (Atlas 1991). Indeed, this is the same approach that is used in mineral processing, and so mine and mill operators readily adapt to the objectives of the treatment. In this paper, passive treatment has been emphasized because it is a cost-effective method for treating effluent from abandoned mines and mills. In passive treatment, maintenance and operations costs are traded for land. All the guidelines given above also could be applied to active systems. In this case, operation and maintenance costs would increase. However, an active system should be able to maintain the activity of bacteria at optimum levels. In active treatment, maintenance and treatment costs would be offset by increased efficiency of treatment.

ACKNOWLEDGMENTS

These studies have been funded by the Superfund Innovative Technology Evaluation (SITE) Program of the U.S. Environmental Protection Agency (E. R. Bates, project supervisor) and the U.S. Bureau of Mines (V. R. Shea-Albin, project supervisor). Steve Machemer and Laura Duggan also have contributed significantly to these projects.

REFERENCES

Atlas, R. M. 1991. "Bioremediation of Fossil Fuel Contaminated Soils." In *In Situ Bioreclamation*, R. E. Hinchee and R. F. Olfenbuttel (Eds.), pp. 14-32, Butterworth-Heinemann, Stoneham, MA.

Bolis, J. L., T. R. Wildeman, and R. R. Cohen. 1991. "The Use of Bench Scale Permeameters for Preliminary Analysis of Metal Removal from Acid Mine Drainage by Wetlands." In W. R. Oaks and J. Bowden (Eds.), *Proceedings of the 1991 National Meeting of the American Society of Surface Mining and Reclamation*, pp. 123-136. American Soc. Surf. Mining and Reclam., Princeton, WV.

Brodie, G. A. 1991. "Achieving Compliance with Staged, Aerobic, Constructed Wetlands to Treat Acid Drainage." In W. R. Oaks and J. Bowden (Eds.), *Proceedings of the 1991 National Meeting of the American Society of Surface Mining and Reclamation*, pp. 151-174. American Soc. Surf. Mining and Reclam., Princeton, WV.

Brodie, G. A., C. R. Britt, T. M. Tomaszewski, and H. N. Taylor. 1991. "Use of Passive Anoxic Limestone Drains to Enhance Performance of Acid Drainage Treatment Wetlands." In W. R. Oaks and J. Bowden (Eds.), *Proceedings of the 1991 National Meeting of the American Society of Surface Mining and Reclamation*, pp. 211-228. American Soc. Surf. Mining and Reclam., Princeton, WV.

Duggan, L. A., T. R. Wildeman, and D. M. Updegraff. 1992. "The Aerobic Removal of Manganese from Mine Drainage by an Algal Mixture Containing Cladophora." In *Proceedings of the 1992 National Meeting of the American Society of Surface Mining and Reclamation*, pp. 241-248. American Soc. Surf. Mining and Reclam., Princeton, WV.

Filas, B. A., and T. R. Wildeman 1992. "The Use of Wetlands for Improving Water Quality to Meet Established Standards." In *Proceedings from Successful Mine Reclamation Conference*, pp. 157-176, Nevada Mining Association, Reno, NV.

Hammer, D. A. (Ed.). 1989. *Proceedings of the International Conference on Constructed Wetlands for Wastewater Treatment*. Lewis Publishing Co., Chelsea, MI.

Hedin, R. S., R. Hammack, and D. Hyman. 1989. "Potential Importance of Sulfate Reduction Processes in Wetlands Constructed to Treat Coal Mine Drainage." In D. A. Hammer (Ed.), *Proceedings of the International Conference on Constructed Wetlands*, pp. 508-514. Lewis Publishing, Chelsea, MI.

Machemer, S. D., and T. R. Wildeman. 1992. "Organic Complexation Compared with Sulfide Precipitation as Metal Removal Process from Acid Mine Drainage in a Constructed Wetland." *Jour. Contaminant Hydrology 9*: 115-131.

Postgate, J. R. 1979. *The Sulphate-Reducing Bacteria*. Cambridge Univ. Press, New York, NY.

Reynolds, J. S., S. D. Machemer, T. R. Wildeman, D. M. Updegraff, and R. R. Cohen. 1991. "Determination of the Rate of Sulfide Production in a Constructed Wetland Receiving Acid Mine Drainage." In W. R. Oaks and J. Bowden (Eds.), *Proceedings of the 1991 National Meeting of the American Society of Surface Mining and Reclamation*, pp. 175-182. American Soc. Surf. Mining and Reclam., Princeton, WV.

Stumm, W., and J. J. Morgan. 1981. *Aquatic Chemistry*, 2nd ed. John Wiley & Sons, New York, NY.

Wetzel, R. G. 1983. *Limnology*, 2nd ed. Saunders College Pub., Philadelphia, PA.

Wildeman, T. R. 1992. "Constructed Wetlands that Emphasize Sulfate Reduction: A Staged Design Process and Operation in Cold Climates." In *Proceedings of 24th Annual Operator's Conference of the Canadian Mineral Processors*, Paper 32. Canadian Institute of Mining, Metallurgy, and Petroleum, Ottawa, Ontario, Canada.

Wildeman, T. R., Brodie, G. A., and J. J. Gusek. 1992. *Wetland Design for Mining Operations*, Bitech Publishing Co., Vancouver, BC, Canada.

Wildeman, T. R. and L. S. Laudon. 1989. "The Use of Wetlands for Treatment of Environmental Problems in Mining: Non-Coal Mining Applications." In D. A. Hammer (Ed.), *Proceedings of the International Conference on Constructed Wetlands*, pp. 221-231. Lewis Publishing, Chelsea, MI.

BACTERIAL CHROMATE REDUCTION AND PRODUCT CHARACTERIZATION

R. J. Mehlhorn, B. B. Buchanan, and T. Leighton

ABSTRACT

Bacillus subtilis reduced hexavalent chromate to trivalent chromium under either aerobic or anaerobic conditions. Reduction of Cr(VI) and the appearance of extracellular Cr(III) were demonstrated by electron spin resonance (ESR) and spectrophotometry. Chromate reduction was stimulated more than 5-fold by freeze-thawing, indicating that intracellular reductases or chemical reductants reduce chromate more rapidly than do intact cells. Moderately concentrated cells (10% pellet volume after centrifugation) reduced approximately 40 µM chromate/min (2 mg Cr/L·min) when exposed to 100 µM chromate (5 mg Cr/L). Highly concentrated cells (70% pellet volume) reduced more than 99.8% of 2 mM chromate (100 mg Cr/L) within 15 min. This rate of chromate reduction was of the same order of magnitude as the rate of respiration in aerobic cells. A substantial fraction of the reduction product (ca 75%) was extracellular Cr(III), which could be readily separated from the cells by centrifugation. At high chromate concentrations, some fraction of reduced Cr(VI) appeared to be taken up by cells, consistent with a detection of intracellular paramagnetic products. At low chromate concentrations, undefined growth medium alone reduced Cr(VI), but at a slow rate relative to cells. Under appropriate conditions, *B. subtilis* appears to be an organism of choice for detoxifying chromate-contaminated soil and water.

INTRODUCTION

Chromate, i.e., hexavalent chromium (Cr(VI)), is an established human carcinogen, whose modes of action may include a catalysis of free radical reactions and cross-linking of DNA (Shi et al. 1991). Cr(VI) is generally a potent microbial toxin, although examples of resistant microbes are known. The stable reduction product of chromate, the chromic Cr(III) species, generally shows considerably less toxicity. Several aerobic and anaerobic microorganisms are known to reduce chromate (Ishibashi et al. 1990). It has been suggested that such organisms may have utility in environmental restoration. Here we show that *Bacillus subtilis*, a widely distributed aerobic soil bacterium posing no known hazards to humans and capable of growing under a variety of conditions, is a potent chromate-reducing, and hence, detoxifying, organism.

MATERIALS AND METHODS

ESR Assays

Although chemical species-specific chromium assays with indicator dyes have been applied to the study of microbial chromate metabolism (Ishibashi et al. 1990), these methods do not lend themselves to studies of high chromate exposure levels, where uncharacterized complexes of metabolic products, e.g., complexed Cr(III) species, may interfere with the absorbance measurements. To overcome this potential problem and to provide tools for analyzing the localization of chromate metabolism we developed and applied to *B. subtilis* a variety of novel ESR assays. Included was an assay for chromate ions, capable of detecting as little as 5 µg Cr/L in buffer solutions, described in more detail below. We also developed a less sensitive assay for chromic ions (sensitivity about 10 mg Cr/L), assays for both intracellular and extracellular univalent reductase activities, and an assay for intracellular paramagnetic species, suitable for detecting the accumulation of chromic ion complexes. These assays were supplemented by previously established ESR methods for measuring cell volumes, spectrophotometric analyses of chromate and EDTA-chelated Cr(III), as well as X-ray fluorescence spectroscopy (XRFS) measurements of total chromium and other trace elements.

Cell Preparations

Cells grown to an optical density of 125 Klett units in an enriched medium (modified from Pipper et al., 1977, by dilution with 1 vol deionized water) were packed by centrifugation, resuspended in about 10 volumes of fresh growth medium (referred to as 10% pellet volume cells), and used for all experiments except where an additional centrifugation step was employed to increase cell density. After harvesting, the cells were maintained on ice for up to 1 hour before transfer to the ESR laboratory. Thereafter the cells were stirred on ice. Typically, 1-mL samples were withdrawn from the stirred suspension with a micropipette and treated with chromate or other reagents in a 1.5-mL Eppendorf tube.

ESR Measurement of Cr(VI)

The Cr(VI) assay consisted of analyzing a transient paramagnetic Cr(V)-complex that arises from the univalent reduction of Cr(VI) by thioglycerol and the complexing of Cr(V) by glycerol and thioglycerol. Two reaction cocktails, RC1 to control pH and to facilitate ligand exchange, and RC2 to reduce Cr(VI) and to form a persistent Cr(V) complex, were developed empirically to avoid a suppression of the ESR species by reaction intermediates and products. RC1 consists of 0.5 M $NaHCO_3$, 0.1 M sodium phosphate, pH 7.4, whereas RC2 consists of 0.4 M glycerol and 20 mM thioglycerol in 10 mM sodium phosphate, pH 7.4. The assay consisted of treating 25 µL of an unknown sample with 0.5 µL of 10 mM sodium ferric-EDTA, 5 µL of RC1, and 20 µL of RC2 and placing the mixture into a 100-µL glass capillary for ESR analysis. The ESR signal arising from the reaction was scanned at 2-min intervals to determine its maximum magnitude. Quantification of chromate was relative to ESR signals arising from defined additions of potassium chromate to the unknown solution (recovery experiment). In some experiments, larger ESR signals were obtained by replacing thioglycerol in RC1 by the reduced nitroxide TOLH (1,4-dihydroxy-2,2,6,6-tetramethyl piperidine).

RESULTS AND DISCUSSION

Effect of Chromate on Growth of *B. subtilis* and Elimination of Chromate from Aerobic Growth Media

Cultures of *B. subtilis* grew normally in the presence of 50 µM K_2CrO_4 (2.5 mg Cr/L) and suffered only modest growth inhibition in the presence of 0.5 mM K_2CrO_4 (Figure 1). ESR analyses of supernatant fractions from centrifuged cells grown with 50 µM K_2CrO_4 indicated that more than half of the chromate had been removed from the batch culture medium during the logarithmic phase of growth. The supernatant fractions of cells grown for 2 hours with 50 µM K_2CrO_4 contained 18 µM residual Cr(VI) (1 mg Cr/L). There was no detectable loss of Cr(VI) in parallel samples of uninoculated growth media. Supernatant fractions of cells grown to stationary phase with 25 µM K_2CrO_4 (3 hours of growth) contained less than 0.15 µM chromate (8 µg Cr/L). Parallel samples of uninoculated

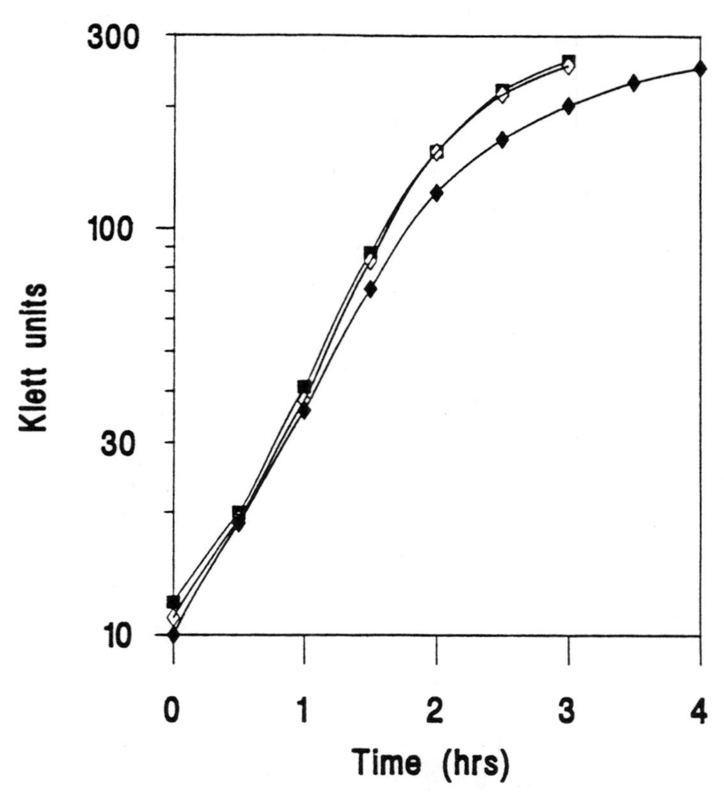

FIGURE 1. Growth curves of *B. subtilis* in the absence and presence of chromate. Control (▪), 50 μM K_2CrO_4 (2.5 mg Cr/L;♦) and 0.5 mM K_2CrO_4 (25 mg Cr/L;◊).

growth media contained 15 μM residual chromate. Thus reduction of low chromate concentrations by undefined growth medium can sometimes be detectable and should be analyzed separately to accurately quantify cellular chromate reduction rates at low chromate levels (ca 1 mg Cr/L). The variability in chromate reduction by growth media may be due to varying degrees of autooxidation of reduced thiols during preparation of the media. However, reduction by medium alone was always substantially less than reduction by cells. At 0.5 mM K_2CrO_4, the chromate concentration in cell supernatant fractions did not change appreciably during the 2-hour logarithmic growth period. Residual chromate was quantified by recovery experiments (linear extrapolation of ESR signals), because the magnitudes of the ESR signals were affected by changes in the medium during cell growth. Analysis of total chromium concentrations by XRFS after centrifugation indicated that the chromium concentration in cell pellets was about the same as the initial chromium concentration in the medium (Table 1, after correcting for the effects of lyophilization and allowing for variability in pellet density). Washing removed much of this chromium, indicating that the bulk of the reduction product during logarithmic growth (low cell density) was present in the extracellular domain. Because the cells were grown under aerobic conditions, these results indicate that chromate detoxification by *B. subtilis* is feasible in aerobic environments.

Chromate Reduction by Anaerobic Suspensions of Concentrated Cells

To gain insight into the potential extent of chromate reduction, we studied concentrated cells by ESR. Cells that had been stirred on ice for about 6 hours after harvest (500 µL) were pelleted by 2 min centrifugation at 10,000 g, diluted with 200 µL of growth medium (70% cell pellet), and treated with 2 mM K_2CrO_4 (100 mg Cr/L) for 15 min. The cells were then diluted 2-fold with 50 mM $MgCl_2$, 10 mM TRIS-HCl, pH 7.0, centrifuged 2 min at 10,000 g and the supernatant solution was assayed for chromate. ESR analysis of the residual chromate showed that these cells had removed

TABLE 1. Trace elements (mg/g dry weight) of pelleted *Bacillus subtilis* grown in chromate solutions.[a]

Element	Control (5/28/92)	2.5 mg Cr/L chromate grown	2.5 mg Cr/L chromate & wash	Control (6/2/92)	2.5 mg Cr/L chromate grown	25 mg Cr/L chromate grown
Ti	8±4	<12	13±4	<12	<12	<12
Cr	<5	53±3	3.7±1.4	2.8±1.4	27±2	1010±50
Mn	33±2	32±2	33±2	33±2	35±2	40±3
Fe	48±2	42±2	44±2	112±6	33±2	33±2
Ni	1.3±0.3	0.7±0.3	<0.9	<0.9	<0.9	0.4±0.3
Cu	8.2±0.5	6.9±0.3	5.1±0.3	9.7±0.5	5.5±0.3	4.5±0.3
Zn	66±3	68±3	61±3	77±4	84±4	95±4
Se	0.3±0.1	0.4±0.1	0.7±0.1	0.3±0.1	0.4±0.1	0.5±0.1
Br	2.1±0.1	2.0±0.1	6.6±0.3	0.4±0.1	0.4±0.1	0.2±0.1
Rb	13.3±0.7	12.5±0.6	12.7±0.6	24±1	16±1	22±1
Sr	3.4±0.3	3.9±0.3	5.0±0.3	3.0±0.3	3.9±0.3	3.8±0.3
Pb	<2	<2	<2	0.7±0.5	<2	0.6±0.5

(a) Trace elements were determined by X-ray fluorescence. Columns 2-4 and columns 5-7 refer to two separate experiments. "Wash" refers to centrifuged cells after resuspension in chromate-free medium.

more than 99.8% of the chromate from the extracellular aqueous phase within 15 min (Figure 2). The supernatant solution was treated with the thioglycerol reaction mixture (see Materials and Methods) and treated with K_2CrO_4 (recovery experiment) yielding 0, 1, 2, or 3 µM of freshly added chromate (0, 50, 100, or 150 µg Cr/L), as indicated. The measured g-value corresponds to that observed when chromate and thioglycerol are mixed in phosphate buffer at pH 7. The final chromate concentration in the supernatant fraction, designated by "0", was below detectability; therefore the residual chromate levels may have been substantially lower than the level shown in sample "1", which had been treated with 1 µM (50 µg Cr/L) fresh chromate. Visual observation supported the ESR assays, i.e., the yellow color seen in the ESR capillaries containing freshly chromate-treated cell suspensions had faded over a period of several minutes and assumed a grayish hue when chromate became undetectable by ESR. Polarographic measurements showed that respiratory activity in these cells consumed dissolved oxygen in less than 1 min; thus most of the chromate removal occurred under anaerobic conditions. This finding establishes the feasibility of chromate removal with anaerobic *B. subtilis*. Visual observation also indicated that the rate of chromate

FIGURE 2. ESR recovery experiment (numbers refer to mM of added chromate), demonstrating the absence of detectable chromate in an aqueous solution, initially containing 2 mM chromate, after 15 min incubation with concentrated *B. subtilis*. The magnetic field interval is shown in millitesla (mT).

fading increased as a function of incubation time of the cell suspension on ice.

Extracellular Nitroxide Reductase Activity: Inhibition by Chromate

The extensive chromate reduction seen in *B. subtilis* was surprising in light of the well-documented toxicity of chromate for many cells. This finding suggested that the lack of chromate toxicity might be due to an exclusion of chromate ions by the cells and would imply that chromate was removed by an extracellular reduction system. To test this hypothesis, we applied a nitroxide ESR assay to extracellular univalent reductases. The ESR assay used a membrane-impermeable nitroxide probe, PECU-Glucam, a persistent free radical derivative of glucosamine previously developed for membrane permeability studies with erythrocytes and other cells (Mehlhorn & Packer 1983). Univalent reduction of this probe eliminates its ESR signal. To confirm that this probe did not enter *B. subtilis*, we used the published methodology for quantifying intracellular aqueous volumes (Mehlhorn & Packer 1983). No evidence of an ESR signal of this probe was evident when cells were treated with the extracellular line-broadening agent Mn-EDTA, consistent with the expectation that this probe is impermeable to membranes of *B. subtilis*.

Supernatant solutions from centrifuged cells (pelleted within 3 hours of harvesting) did not reduce the nitroxide appreciably, indicating negligible nitroxide reductase activity by growth medium or by substances released from cells. In cell suspensions not supplemented with chromate, the nitroxide was reduced at a rate of 7.6 µM/min, corresponding to about 6% of the rate of respiratory oxygen consumption. In the tenth-millimolar concentration range of chromate, nitroxide reduction was transiently inhibited; then nitroxide was reduced at the same rate as had been observed in the absence of chromate (Figure 3A). If the nitroxide was incubated with cells in the absence of chromate until the ESR signal had been eliminated, the signal could be fully restored by treatment with chromate, indicating that the nitroxide reduction product was reoxidized by chromate.

The lag phase in nitroxide reduction was a function of the chromate concentration (Figure 3B), consistent with a rapid reoxidation of reduced nitroxide by chromate until virtually all of the chromate had been depleted. Alternatively, chromate could be preferentially reduced in cell suspensions containing both nitroxide and chromate. By relating the rate of chromate reduction, estimated from the lag phases, to an ESR analysis of cell volumes, a chromate reduction rate of 2.5 nM/min per µL of cell volume was calculated for a chromate concentration of 100 µM (5 mg Cr/L).

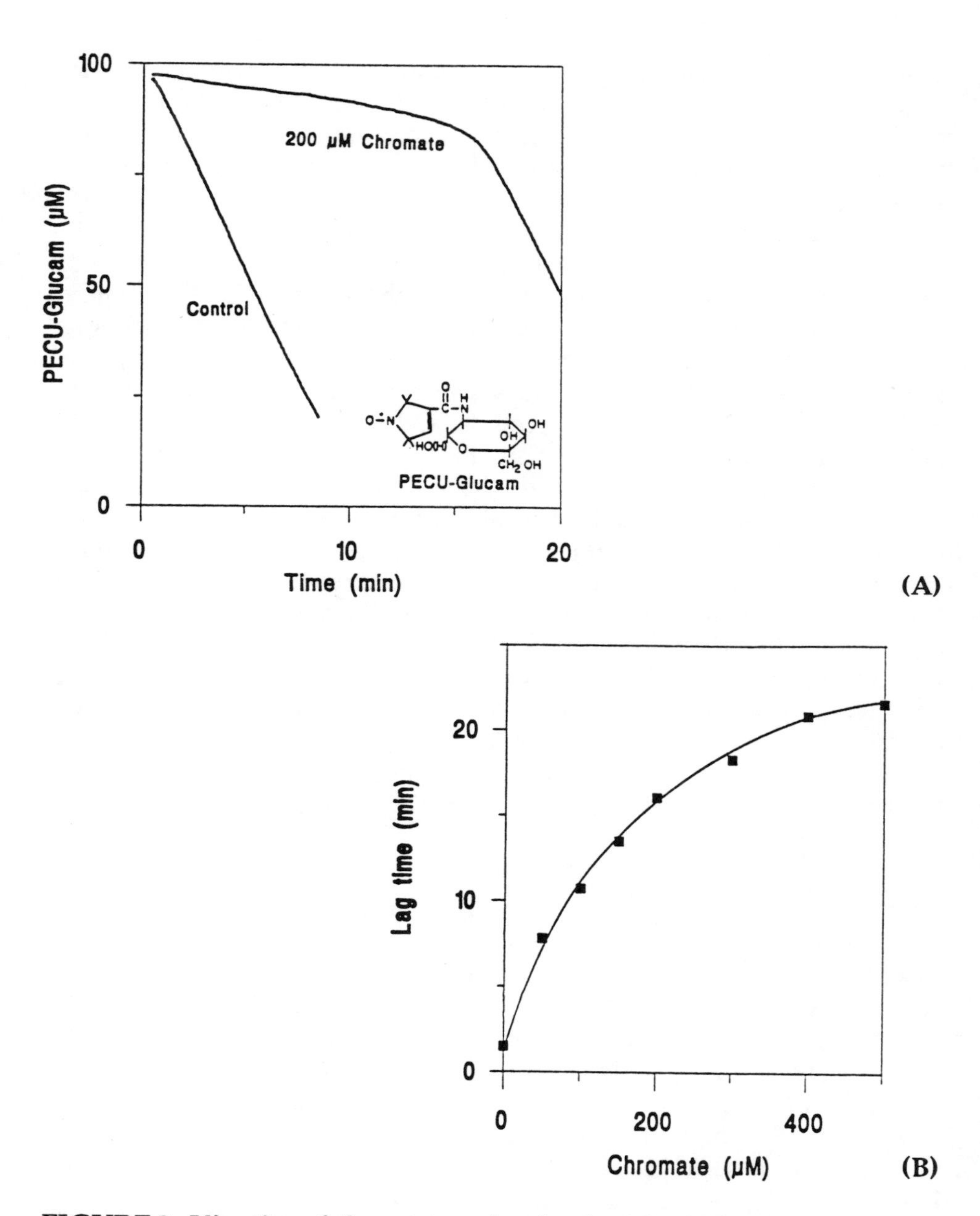

FIGURE 3. Kinetics of chromate reduction by *B. subtilis,* deduced from the inhibition of the reduction of the membrane-impermeable nitroxide PECU-Glucam by chromate. (A) ESR signals of the probe as a function of time showing lag in nitroxide reduction in the presence of 200 mM chromate (10 mg Cr/L); (B) lag phase in nitroxide reduction vs. chromate concentration.

This rate of chromate reduction is comparable to the rate of nitroxide reduction (see also Figure 3A) and one might assume that the nitroxide mediated chromate reduction under these conditions. However, at higher chromate concentrations, the chromate reduction rate was significantly greater than the reduction rate for the nitroxide (Figure 3A vs. 3B), suggesting that most of the chromate reduction occurred independently of the nitroxide. The increasing rate of chromate reduction with increasing chromate concentrations (Figure 3B) suggests that the putative reductase activity was not saturated at the lower chromate concentrations.

The lag phase in nitroxide reduction was decreased more than 5-fold when cells were freeze-thawed (three cycles of freezing in liquid nitrogen and thawing at 25°C). Thus intracellular reducing agents have considerably more chromate-reducing activity than does the putative extracellularly oriented reductase.

As noted earlier, virtually complete elimination of 2 mM chromate (100 mg Cr/L) by highly concentrated cells occurred within 15 min, a significantly faster rate of chromate disappearance than the 20-min time frame for chromate reduction in the submillimolar range shown in Figure 3B. The primary reason for this difference is probably the higher cell density of cells used to remove the 2-mM chromate (70% vs. 10% cell pellets). As noted above, aged concentrated cells exhibited higher rates of chromate reduction than cells tested shortly after harvesting; this may have played a role in the extensive chromate reduction observed with the highly concentrated cells (less than 0.2% residual chromate).

Appearance of Extracellular Cr(III)

Although it seemed likely that the mechanism effecting chromate disappearance was an extracellular reduction of Cr(VI) to the stable reduction product Cr(III), direct evidence for the formation of this species was desirable. Because Cr(III) is paramagnetic, it interacts with nitroxides in a concentration-dependent manner. The effect arises from collisions between paramagnetic species and is referred to as spin exchange. It causes the ESR line widths of the nitroxide to become broadened. Previous ESR studies had identified Cr(III) trioxalate as a particularly effective line-broadening agent for nitroxides and we pursued an analytical strategy based on this observation. When stock solutions of $CrCl_3$ were freshly diluted into 0.1 M sodium oxalate containing a ^{15}N perdeuterated nitroxide (characterized by narrow intrinsic line widths and therefore very sensitive to line broadening), a time-dependent line broadening of the nitroxide was observed, with the line width attaining its maximum value in about 10 min. The increase in line width was proportional to the concentration of Cr(III) and was detectable at about 0.5 mM $CrCl_3$ (25 mg Cr/L). We assumed that a time- and chromate-treatment-dependent increase in line widths of the nitroxide after incubation of cell supernatant fractions with sodium oxalate would be indicative of Cr(III). Application of this assay to centrifuged, chromate-treated cells yielded line broadening consistent with a reduction of chromate to Cr(III), of which a substantial fraction remained in the extracellular aqueous phase.

Detection of Intracellular Chromic Ions

ESR volume measurements of chromate-treated cells showed a decrease of the intracellular signal intensities, which was accompanied by an increase in line widths (Figure 4). The maximum magnitude of the line width increase was a function of the chromate concentration, consistent with a conversion of chromate to an intracellular paramagnetic chromium species. It is noteworthy that ESR volumes were not eradicated by any of the tested chromium concentrations, i.e., the observed line broadening (2-fold) was consistent with the line height decreases (4-fold) in the first-derivative ESR spectra. This observation indicates that a substantial fraction of the *B. subtilis* membranes remained intact throughout this chromate treatment.

Spectrophotometric Assays of Cr(VI) Loss and Cr(III) Appearance

The observation of a fading of the yellow chromate color observed in the ESR experiments was pursued with a spectrophotometric study of chromate-treated *B. subtilis* (ca 30% cell pellet volume) by use of a spectrophotometric assay of the EDTA complex of Cr(III). After a 20-min incubation period with 2 mM K_2CrO_4 (100 mg Cr/L), cells were centrifuged and supernatant solutions were analyzed spectrophotometrically. A residual chromate concentration of 250 µM was estimated (absorbance at 350 nm, corrected for absorbance by nonchromate components of growth medium). Thereafter, the supernatant solutions were treated with 20 mM EDTA, pH 7.0. The purple color characteristic of the Cr(III)-EDTA complex developed very slowly in the supernatant, requiring an incubation of 3 days at room temperature to ensure that full color development had occurred. From the absorbance at 550 nm, the Cr(III)-EDTA concentration was estimated to be 1.5 mM. These results support the interpretation that the major species at the beginning and end of the incubation periods were Cr(VI) and Cr(III), respectively. Cr(VI) was

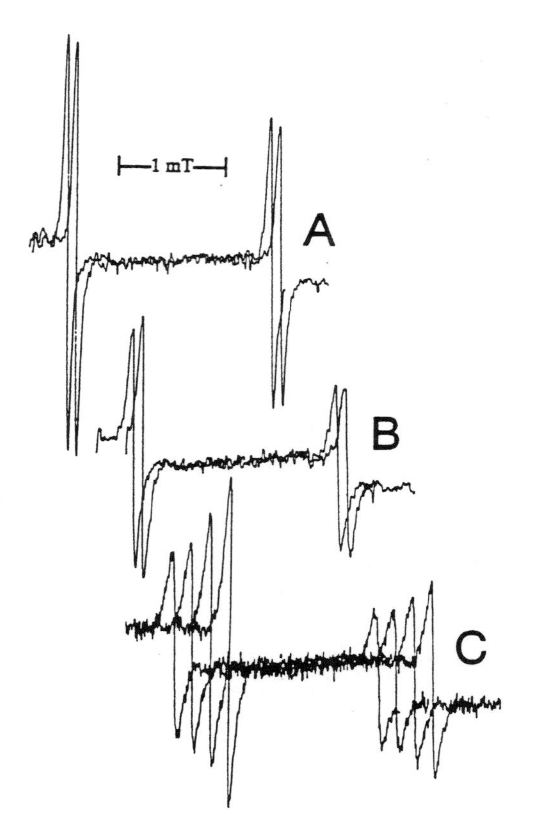

FIGURE 4. Appearance of intracellular paramagnetic species in *B. subtilis* treated with 5 mM K_2CrO_4 (250 mg Cr/L). *B. subtilis* cells (internal aqueous volume of 1.7%) were treated with 100 µM $^{15}N,^{2}H$ Pyrad, 4 mM $K_3Fe(CN)_6$, and 90 mM $(Me_4N)_2$Mn-EDTA. (A) Two scans without chromate, successive spectra shifted to the left, 2 min between scans, instrument gain 2.5×10^4; (B) two scans after treatment with 5 mM K_2CrO_4, instrument gain 2.5×10^4, 2 min between scans; (C) four more scans of chromate-treated samples, instrument gain 2.5×10^4, 4 min between scans.

substantially reduced within 20 min; however, we cannot rule out that Cr(III) formation from an intermediate reduction product may have occurred much more slowly. It can be concluded that the bulk of the chromate reduction product appears in the extracellular phase. This creates the opportunity for developing schemes for chromium immobilization, e.g., using ion exchange matrices (work is in progress).

CONCLUSIONS

Reduction of chromate by the soil bacterium *Bacillus subtilis* has been demonstrated under laboratory conditions. A substantial fraction of the Cr(III) that is produced remains in the extracellular aqueous phase. While *Bacillus subtilis* requires oxygen for growth, its capacity to reduce chromate under both aerobic and anaerobic conditions could be an asset for in situ treatments leading to transient anaerobiosis. If prolonged oxygen depletion were to cause cell death and membrane lysis, our results with freeze-thawed cells suggest that chromate reduction would probably continue. It is likely that chromate reduction would be reversible if Cr(III) were produced in a soluble form. Therefore, it is desirable to produce insoluble Cr(III) species during the reduction of Cr(VI) by *Bacillus subtilis*, perhaps by manipulations of pH and anions.

In summary, reduction of Cr(VI) by *Bacillus subtilis* and precipitation of Cr(III) with appropriate agents to immobilize it and to prevent its reoxidation could be an effective in situ strategy for the long-term detoxification of chromate-polluted environments.

ACKNOWLEDGMENTS

We thank Dr. Ian Fry for assistance with the spectrophotometric analyses, Don Carlson for growing the cells, and Robert Giauque for performing the XRFS analyses. This work was supported by the Office of Technology Development, U.S. Department of Energy and by the Director, Office of Energy Research, Division of University and Science Education Programs of the U.S. Department of Energy under Contract DE-AC-76SF00098.

REFERENCES

Ishibashi, Y., C. Cervantes, and S. Silver. 1990. "Chromium reduction in *Pseudomonas putida*." *Applied and Environmental Microbiology 56*: 2268-2270.

Mehlhorn, R. J., and L. Packer. 1983. "Bioenergetic studies of cells with spin probes." *Proceedings of the New York Academy of Science 414*: 180-189.

Pipper, D. J., C. W. Johns, C. L. Ginther, T. Leighton, and S. Whitman. 1977. "Erythromycin resistant mutations in *Bacillus subtilis* cause temperature sensitive mutations." *Molecular and General Genetics 150*: 147-159.

Shi, X., N. S. Dalal, and V. Vallyathan. 1991. "One-electron reduction of carcinogen chromate by microsomes, mitochondria, and *Escherichia coli*: Identification of Cr(V) and ·OH radical." *Archives of Biochemistry and Biophysics 290*: 381-386.

MICROBIAL REMOVAL OF HEAVY METALS AND SULFATE FROM CONTAMINATED GROUNDWATERS

L. J. Barnes, P. J. M. Scheeren, and C. J. N. Buisman

ABSTRACT

Groundwater beneath well-established metal-refining sites often contains heavy metals and/or sulfate. Increasing demand for potable water from aquifers, coupled with tighter environmental controls, makes it imperative that these contaminants be prevented from spreading. To achieve this goal a geohydrological control system can be installed to extract groundwater; however, this water must be treated before it can be discharged. A sulfate-reducing bacteria process has been developed to remove heavy metals and sulfate from contaminated water. The purified water can then be discharged directly to the environment. The process, which operates under neutral (pH 7) conditions, uses a consortium of naturally occurring anaerobic microorganisms that were isolated from the metal-refining site. The organisms use ethanol as their growth substrate and as energy source for the reduction of sulfate to sulfide. When heavy metals are present, they precipitate as their extremely insoluble sulfides. Heavy metals in solution are lowered to µg/L levels, with sulfate being reduced to less than 200 mg/L. Extended operation of a process demonstration unit proved that the organisms are not affected by wide variations in feed composition, and the process rapidly recovers from operational upsets, and is easy to start up. A commercial-scale plant, with a 1,800 m^3 sulfate-reducing bacteria (SRB) process reactor capable of treating 7,000 m^3/day of extracted groundwater, has been successfully installed and commissioned.

INTRODUCTION

Historically, the metal industry has produced substantial amounts of heavy metals and/or sulfate-containing sludges and solid wastes that often were stored on production sites. In some instances, leaching of the wastes by percolating rain and surface waters has contaminated the groundwater. An example exists at Budel (the Netherlands) where the company Kempensche Zinkmaatschappij was founded in 1892. The original pyrometallurgical zinc plant produced 'zinc ashes', which were used as landfill for plant extensions. At that time these ashes were considered inert. In 1973 a hydrometallurgical zinc plant, capable of producing 200,000 tonnes/year of zinc, was installed. Authorities are now beginning to insist that such contaminated groundwaters be prevented from spreading. Water can be extracted with a geohydrological control system (GCS) by using an array of strategically positioned boreholes, but it must be decontaminated before being discharge to the environment.

In the mid-1980s, Shell Research initiated a program to examine candidate processes capable of removing heavy metals (and sulfate) from extracted groundwater and able to meet discharge criteria required by the authorities. The three most suitable processes selected for research were liquid membrane extraction, ion exchange, and anaerobic microbial sulfate reduction; only the latter will remove sulfate economically.

This paper is an overview of the experimental and process development work carried out to establish the feasibility of the sulfate-reducing bacteria (SRB) process. The GCS has been discussed elsewhere (Scheeren et al. 1991).

This project culminated in designing a groundwater treatment process that was installed at the Budel zinc refinery site in 1992.

The key features of the SRB reactor process are, in situ generation of sulfide by microbes, and concomitant precipitation of heavy metals as highly insoluble sulfides. The development of the SRB process involved two phases:

1. Laboratory investigations (Barnes et al. 1991), including:
   - screening studies to identify suitable mixed microbial cultures,
   - identifying a simple, readily available, economically attractive carbon substrate,
   - establishing the SRB process operating parameters, in continuously operated bench-scale reactors using a synthetic groundwater.
2. Development and process scaleup (Scheeren et al. 1991), including:
   - establishing the process operating window by using an SRB process demonstration unit with a groundwater feed,

- demonstrating an integrated process by using a pilot plant with SRB reactor and downstream treatment of the reactor effluents,
- designing an integrated process compatible with the existing metal-refining site's infrastructure that will meet the required safety and engineering standards.

THE SRB PROCESS

Many natural anaerobic aqueous environments, such as those found in oceans, lakes, and sediments, contain microorganisms that use organic compounds for growth and as the energy source for reducing sulfate to sulfides. When heavy metals are present, they precipitate as extremely insoluble sulfides forming part of the sediment. These organisms play a major role in the formation of some sedimentary metal sulfide deposits as well as producing the iron sulfides found in coal (Krouse & McCready 1979). The overall reaction in such an anaerobic environment can be represented by:

$$\text{Metal sulfate} + \text{Carbon substrate} \rightarrow$$
$$\text{Metal sulfide} + CO_2 + H_2O + \text{cells}$$

SRB normally occur as part of a consortium of interdependent anaerobic organisms. Such a consortium exists because of the presence of a complex mixture of naturally occurring carbon compounds. Because SRB use a limited range of simple organic substrates, they take advantage of products excreted by primary organic-degrading organisms. To sustain active growth of SRB, neutral (pH 7), reductive conditions must be maintained (Brown et al. 1973). Such conditions can be attained in an anaerobic bioreactor.

It should be noted that the quantity of sulfate reduced is proportional to the growth of the SRB, which in turn is related to the energy available in the organic growth substrate. Sulfate conversion can therefore be controlled by varying the amount of carbon substrate supplied to the culture.

LABORATORY INVESTIGATIONS

The experimental program, which started in 1987, initially involved isolating SRB from a wide range of environmental samples taken from; oil- and metal-refining sites, fresh and saline rivers, fresh water canals, and sewage works. All the SRB active cultures were combined to give a mixed culture that was subsequently used in laboratory reactor studies to ascertain conditions for optimal growth on an economically attractive carbon substrate. Ethanol was found to be a suitable substrate, however, it is a major contributor to process operating costs.

Experimental details are published elsewhere (Barnes et al. 1991).

Results and Discussion

Carbon Growth/Energy Source. Ethanol is an ideal carbon substrate for SRB, but small quantities of acetate are formed as a by-product. However, the addition of methanogens to the culture ensures complete degradation of the acetate; thus producing a low biological oxygen demand effluent. Steady-state data from reactor experiments showed that approximately 1 mol of sulfate is reduced per mol of ethanol consumed (i.e., 96 mg/L sulfate reduced per 46 mg/L ethanol consumed); as indicated in Figure 1. The overall reaction can be represented by

$$C_2H_5OH + SO_4^{2-} \rightarrow S^{2-} + 1.35CO_2 + 2H_2O + 0.35CH_4 + 0.3CH_2O \text{ (cells)}$$

Nutrients. Ammonium salts or urea are ideal nitrogen sources, and phosphate is the most suitable source of phosphorus. All other trace elements are present in the groundwater. Complete degradation of ethanol to carbon dioxide, methane, and biomass is achieved, providing that the molar ratios for ethanol/nitrogen and ethanol/phosphorus are less than 20 and 500, respectively.

pH Requirement. The optimum pH for biomass growth and sulfate reduction was shown to be 7.5; however, these organisms are active in the pH range of 6 to 8 and will survive excursions to pH 4 and pH 9.

Temperature and Liquid Residence Time. The maximum operating temperature is defined by the organisms' tolerance, being 42°C for our system. The minimum liquid residence time required in the SRB reactor to achieve the desired sulfate reduction, solids retention, and complete degradation of carbon substrate has been shown to be dependent on the settling velocity of the bed particles and associated organisms and not on reaction velocities or organism growth. However, below 15°C the minimum liquid residence time is determined by the ethanol/acetate degradation rate.

Effect of Redox Potential. A low redox potential is essential to maintain a stable culture and maximize sulfate reduction. A maximum

UNIT		Laboratory		Demonstration		Demonstration		Pilot Plant		Commercial		
Reactor Volume	m^3	0.002		9		9		12		1800		
Residence Time	hour	4		8		11		5		7		
Temperature	°C	30		34		21		35		35		
		FEED	OUTLET	FEED	OUTLET	FEED	OUTLET	FEED	OUTLET	FEED	OUTLET	TARGET
pH		3.2	7.0	6.7	7.2	4.9	7.0	5.0	6.9	5.5	7.5	6.5–9
Redox Potential	mV	nd	−300	+20	−360	+190	−300	+180	−295	nd	nd	–
Ethanol	mg/L	585	<1	360	<1	1580	<1	730	<1	120	<1	<1
Sulfate	mg/L	1480	180	805	165	3200	195	1720	150	380	<200	<200
Zinc	mg/L	112	<0.004	107	0.17	1070	0.22	367	<0.05	50	<0.3	<0.3
Cadmium	mg/L	3.47	<0.006	0.87	<0.01	18	<0.01	6.4	<0.01	0.1	<0.01	<0.01
Cobalt	mg/L	2.05	<0.019	0.14	<0.02	0.51	<0.02	0.26	<0.01	0.1	<0.03	<0.03
Copper	mg/L	2.98	0.009	0.46	<0.02	6.8	<0.02	2.26	<0.01	0.1	<0.02	<0.02
Iron	mg/L	6.08	0.062	49	0.03	67	0.01	60	0.05	80	<1	<1
Calcium	mg/L	315	237	360	365	370	392	348	330	<100	<100	<300
Magnesium	mg/L	27.5	27.3	17.4	18.5	140	116	56	46	<50	<50	<50

FIGURE 1. Typical performance data from laboratory reactor, process development unit, pilot plant, and commercial process.

redox potential of −100 mV is sufficient to sustain SRB activity (Brown et al. 1973); however, a redox close to −300 mV is required for the methanogens to maintain optimum activity. Such a redox is given by the presence of soluble sulfide (Pourbaix 1963). However, short excursions to high redox have no detrimental effect on the organisms, but recovery times are relatively long unless soluble sulfide is added to lower the redox potential.

Heavy Metal Removal. Efficient removal of heavy metals is the most important attribute of the SRB process. Providing excess soluble sulfide is present, the heavy metal concentrations in solution depend on the solubility products of the metal sulfides. However, the total metal concentration in the process effluent is determined by flocculation of the sulfide particles. To reach µg/L levels of heavy metals, subsequent filtration is required.

Inhibitory Effects of Feed Components. Inhibitory effects of potential feed components on microbial growth and hence sulfate reduction must be considered. As expected, alkali and alkaline earth cations have no deleterious effect, because the organisms used can be isolated from marine environments (Widdel & Pfennig 1984). Heavy metal cations are nontoxic, presumably because the presence of soluble sulfide maintains a very low metal concentration in solution. Certain anions are, in general, known to be powerful microbial inhibitors. However, the following anions were found not to inhibit microbial growth or sulfate reduction at the concentrations tested (mg/L): sulfide (400), selenate (3.5), arsenate (3.3), fluoride (48), and molybdate (10).

Mass Balance over the SRB Reactor. The sludge blanket in the reactor can be considered to be in perfect equilibrium with the feed and effluents only after a long period of operation (months), providing the feed composition and reactor conditions have remained constant during the whole period. However, a sensible and useful overall mass balance can be attained from analysis of the feed, outflow liquid, and gas phases, assuming the elemental difference is associated with the sludge. Figure 2 shows a typical mass balance over an SRB reactor operated under such a pseudosteady-state condition.

DEVELOPMENT AND PROCESS SCALEUP

For purification of groundwater extracted from the Budel zinc-refining site, it was necessary to develop a commercial-scale water treatment plant capable of handling 7,000 m^3/day of extracted groundwater and producing an environmentally acceptable effluent.

Process Development Unit (PDU)

A PDU (shown schematically in Figure 3), with a 9-m^3 upflow sludge blanket reactor, was operated with groundwater feed to demonstrate scaleup of the SRB process from a 1.5-L bench-scale reactor. Operation of the PDU showed that a flocculant was required to obtain low metal sulfides and biomass carryover in the aqueous effluent at an acceptable liquid residence time (less than 8 hours). Extensive trials with this unit proved that the organisms are not affected by wide variations in feed composition, and the process rapidly recovers from operational upsets, and is easy to start up. Typical data are shown in Figure 1.

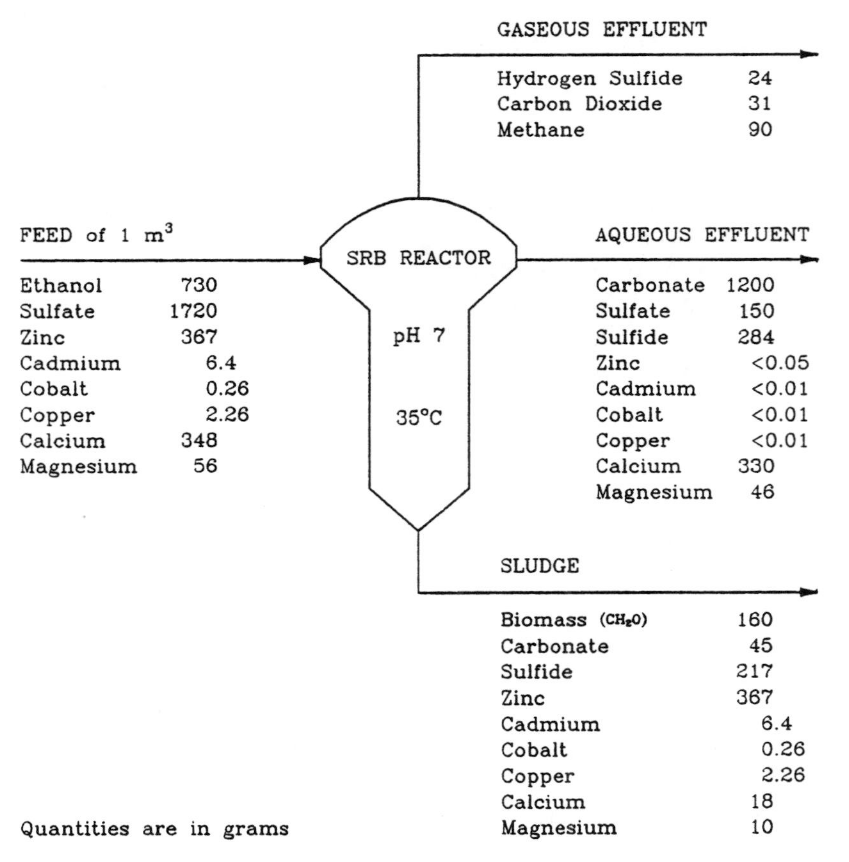

FIGURE 2. Typical mass balance over an SRB reactor.

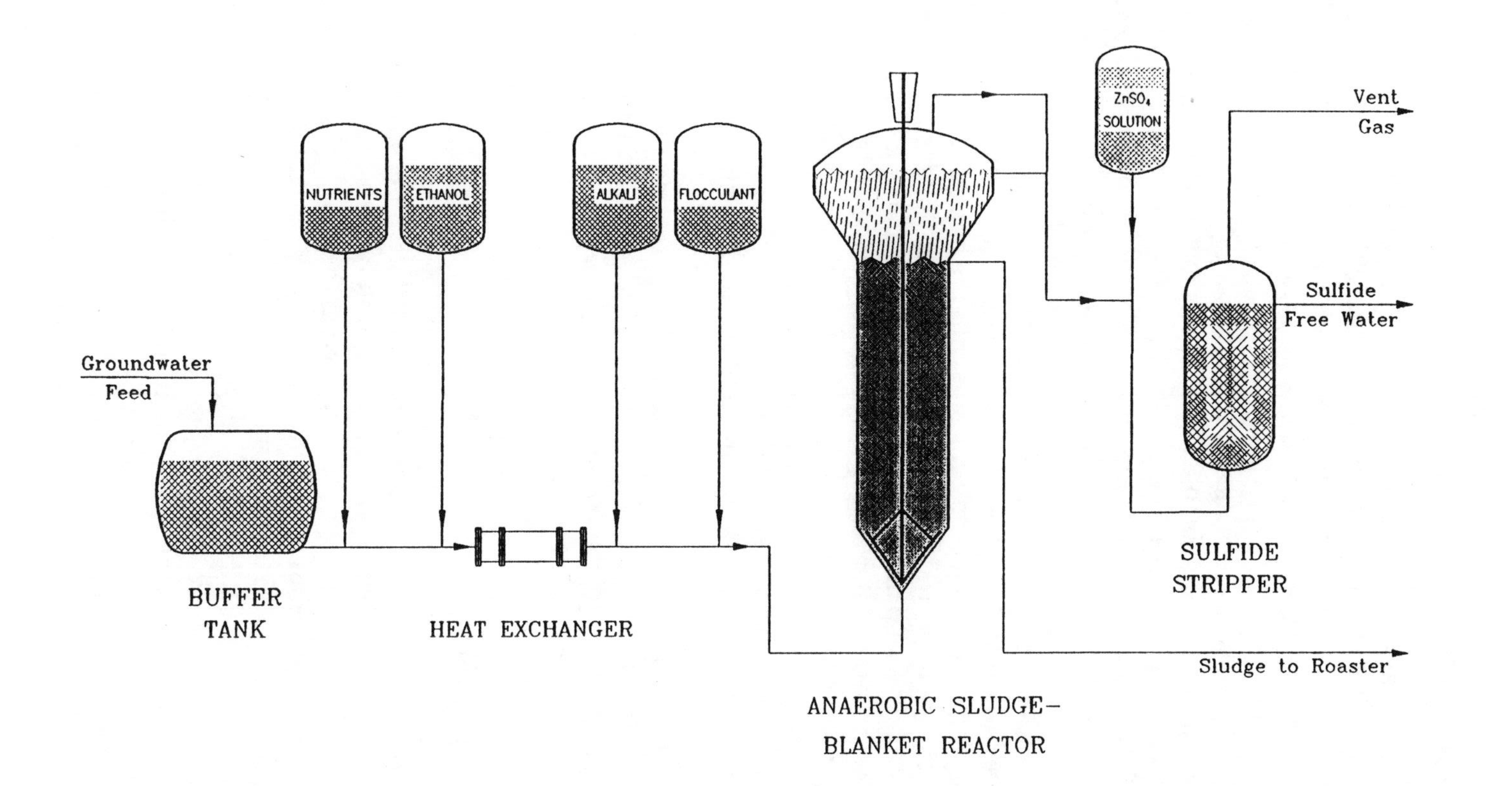

FIGURE 3. Schematic diagram of the process development unit.

Pilot Plant (PP)

Demonstration of a completely integrated process was necessary because the SRB reactor produces an aqueous effluent, and a sludge, both of which contain soluble hydrogen sulfide, as well as a gaseous effluent, which is predominately methane containing carbon dioxide and hydrogen sulfide.

The PP, shown schematically in Figure 4, consisted of a 12-m^3 SRB upflow anaerobic sludge-blanket (UASB) reactor, a submerged fixed-film (SFF) reactor, a tilted plate settler, and a polishing sand bed filter. In the SFF reactor, aerobic microorganisms (*Thiobacilli*) convert soluble sulfide to sulfur (Buisman et al. 1990). The sulfur from the SFF reactor and solids carryover from the UASB reactor are removed by a settler and filter. For gas treatment, a gas scrubber and a flare were installed. The PP is described in detail by Scheeren et al. (1991). This plant was used for 8 months at the Budel zinc-refining site with the objective of determining the efficiency of each process operation and fix the design parameters for the commercial plant. Typical data are shown in Figure 1. In addition, the PP was used to determine the consistency of operation and an optimal startup procedure.

This integrated water treatment process produces an environmentally acceptable aqueous effluent for direct discharge and a sludge containing heavy metals and sulfur that can be recovered by mixing with the zinc refinery's concentrate feed.

A full-size integrated process capable of treating 7,000 m^3/day of groundwater has been designed and installed at the Budel zinc-refining site. Typical data for the commercial plant, which was commissioned in November 1992, are shown in Figure 1.

CONCLUSIONS

Tighter environmental control within Western Europe requires that heavy metal and sulfate-contaminated groundwaters should be treated. Adherence to the regulations has resulted in the development of a heavy metal and sulfur removal/recovery technology based on the capability of bacteria to reduce sulfate in an anaerobic environment.

This anaerobic SRB process can remove a wide concentration range of heavy metals from aqueous streams, giving an aqueous effluent containing only a few µg/L of the metals. In addition, sulfate can be simultaneously lowered to less than 200 mg/L.

SRB consume about 1 mol of ethanol (the carbon/energy substrate) in reducing 1 mol of sulfate and produce a small quantity of acetate as

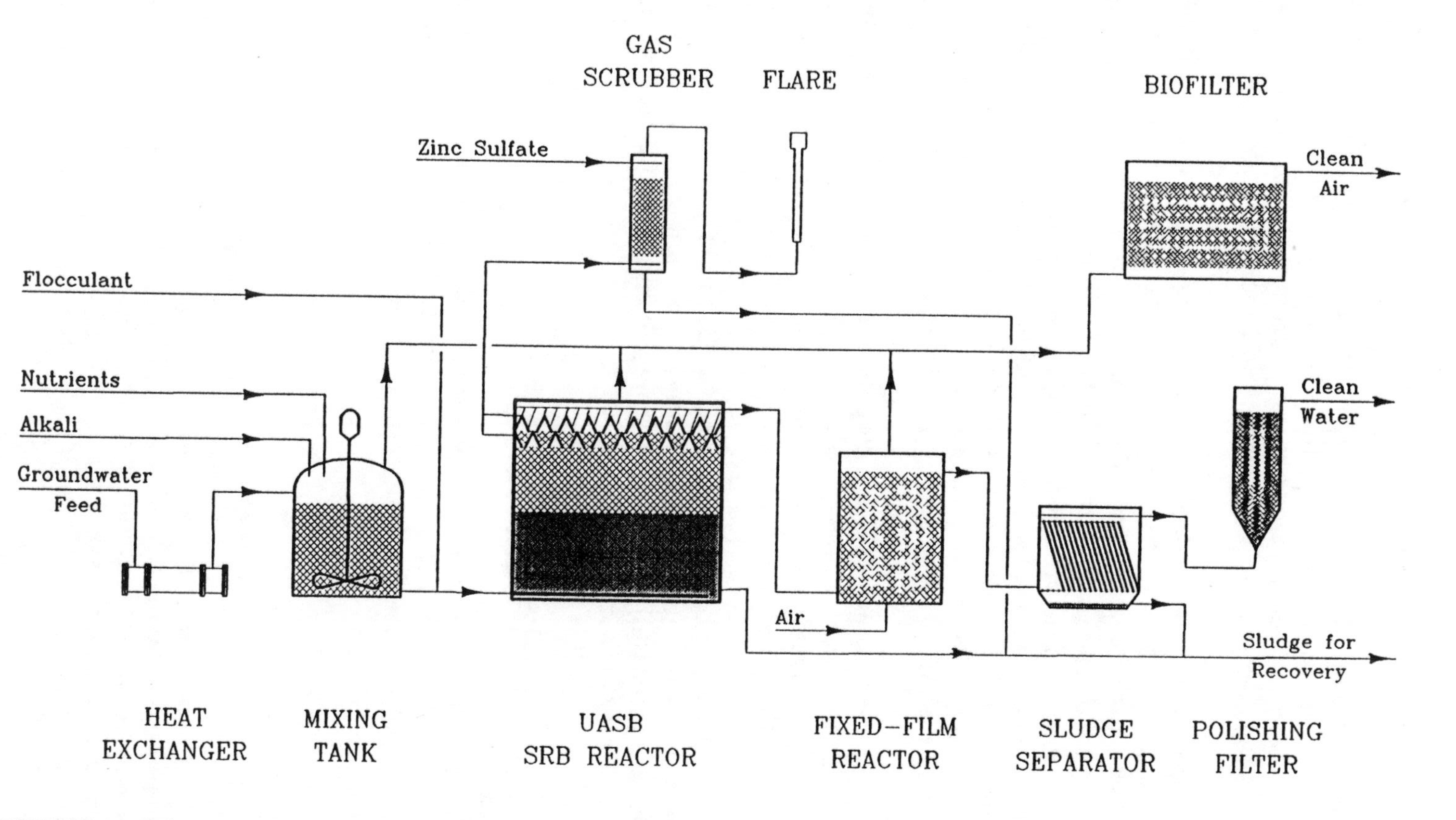

FIGURE 4. Schematic diagram of the pilot/commercial plant.

a by-product. The acetate can be successfully degraded to carbon dioxide, methane, and biomass by ensuring that methanogens coexist in the SRB culture. To sustain microbial growth and produce an effluent with a low biological oxygen demand, nitrogen and phosphorus levels in the reactor feed must be greater than 0.05 mol and 0.002 mol, respectively, for each mol of ethanol consumed.

Temperature has a negligible effect on the performance of the SRB reactor in the range 15 to 40°C. Addition of flocculant to the SRB reactor feed is essential if a short liquid residence time (less than 8 hours) and adequate solid retention are to be achieved within a single reactor.

Unwanted sulfide present in the aqueous product stream from the SRB reactor is converted to sulfur by controlled biochemical oxidation; the sulfur and solids are subsequent removal before discharge. The biogas is flared after the hydrogen sulfide has been scrubbed out. The biosludge, consisting mainly of metal sulfides, can be treated in a metal-refining process.

The anaerobic mixed culture employed is hardy, can handle many potentially inhibitory cations and anions, and recovers readily from (substantial) process upsets. The complete integrated process is easy to start up and requires minimal attention and maintenance.

This successful research and development program has culminated in the installation of a commercial process. Commissioning trials demonstrated that the groundwater at the Budel zinc-refining site can be purified to give an aqueous effluent that is environmentally acceptable and satisfies the regulatory authorities.

ACKNOWLEDGMENTS

Close cooperation between Shell's Sittingbourne Research Centre, Budelco B.V. and PAQUES Environmental Technology, has played an important role in the successful implementation of this water treatment process.

REFERENCES

Barnes, L. J., J. Sherren, F. J. Janssen, P. J. M. Scheeren, J. H. Versteegh, and R. O. Koch. 1991. "Simultaneous microbial removal of sulfate and heavy metals from waste water." In *Proceedings of EMC'91: Non-Ferrous Metallurgy – Present and Future*, pp. 391-402. Elsevier Applied Science, London, UK.

Brown, D. E., G. R. Groves, and J. D. A. Miller. 1973. "pH and Eh control of cultures of sulfate-reducing bacteria." *Journal of Applied Chemical Biotechnology* 23:141-149.

Buisman, C. J. N., B. G. Geraats, P. Ijspeert, and G. Lettinga. 1990. "Optimization of sulfur production in a biotechnological sulfide-removing reactor." *Biotechnology Bioengineering* 35(1):50-56.

Krouse, H. R., and R. G. L. McCready. 1979. "Biogeochemical cycling of sulfur." In P. A. Trudinger, and D. J. Swaine (Eds.), *Biogeochemical Cycling of Mineral-Forming Elements*, pp. 401-430. Elsevier Scientific Publishing Co., New York, NY.

Pourbaix, M. 1963. *Atlas of Electrochemical Equilibria in Aqueous Solution*, National Association of Corrosion Engineers, Houston, TX.

Scheeren, P. J. M., R. O. Koch, C. J. N. Buisman, L. J. Barnes, and J. H. Versteegh. 1991. "New biological treatment plant for heavy metal contaminated groundwater." In *Proceedings of EMC'91: Non-Ferrous Metallurgy – Present and Future*, pp. 403-416. Elsevier Applied Science, London, UK.

Widdel F., and N. Pfennig. 1984. "Dissimilatory sulfate- or sulfur-reducing bacteria." In N. R. Krieg, and J. G. Holt (Eds.). *Bergey's Manual of Systematic Bacteriology*. Volume 1, pp. 663-679. Williams & Wilkins, Baltimore, MD.

COMBINED REMOVAL OF ARSENIC, VOCS, AND SVOCS FROM GROUNDWATER USING AN ANAEROBIC/AEROBIC BIOREACTOR

D. S. Lipton, J. M. Thomas, G. M. Leong, and K. Y. Henry

ABSTRACT

Groundwater at a coatings manufacturing facility contained about 60 mg/L total nonchlorinated volatile organic compounds (VOCs), 1 mg/L total nonchlorinated semi-VOCs (SVOCs), and up to 150 mg/L total arsenic. A laboratory study was conducted to evaluate a biological process for treating VOCs and SVOCs and for reducing the concentrations of arsenic in the aqueous phase. The biotreatability study was conducted using a four-cell, continuous-flow bioreactor that could be operated under anaerobic (Cells 1 and 2) and aerobic (Cells 3 and 4) conditions. The results of the biotreatability study demonstrated removal of organic compounds to their analytical detection limits (0.005 to 0.02 mg/L) under both an anaerobic/aerobic configuration and an aerobic-only process. No detectable amounts of VOCs or SVOCs were measured in the vent gas. The total amount of volatile arsenic measured in the headspace of the anaerobic cells and in the vent gas from the aerobic cells accounted for less than 0.01% of the total mass of arsenic fed into the bioreactor. Arsenic concentrations in the groundwater remained unchanged when the bioreactor was operated under only aerobic conditions. However, under combined anaerobic/aerobic operating conditions, arsenic concentrations in the groundwater were reduced by more than 70% in the first anaerobic cell and by another 50% in the second anaerobic cell. The addition of $FeCl_3$ to the second aerobic cell increased the total removal of arsenic up to 99.7%, probably through precipitation of the most oxidized form of arsenic (arsenate) with iron. The lowest effluent concentration of arsenic achieved by this process was 0.09 mg/L. Biomass collected from Cells 1, 2, 3, and 4 was found to contain about 30%, 25%, 7%, and 2% arsenic on a dry weight basis, respectively. The anaerobic/aerobic biological process evaluated in this treatability study demonstrated the potential for treating groundwater containing a mixture of organic compounds and metalloids such as arsenic.

INTRODUCTION

A laboratory study was undertaken to evaluate the potential for treating groundwater from a coatings manufacturing facility ("the Site") that contained arsenic (greater than 150 mg/L), nonchlorinated volatile and semivolatile organic compounds (VOCs and SVOCs, respectively), using a submerged, fixed-film, continuous-flow biological treatment system ("the bioreactor"). The treatment of nonchlorinated VOCs and SVOCs using biological processes is well established. However, reports describing successful biotreatment in the presence of large amounts of soluble arsenic and its concurrent removal in biological treatment systems are not known to the authors. The potential for a biological treatment process to remove arsenic is based on its chemistry, which provides the possibilities for arsenic's removal as elemental arsenic and arsenide under reducing conditions, and as precipitates of iron hydroxides under oxic conditions (Ghosh 1986; O'Niell 1990; Korte & Fernando 1991).

Preliminary respirometry studies established that groundwater from the Site was not toxic to aerobic microorganisms. Therefore, the objectives of this biotreatment study were to evaluate (1) the potential for biotreating VOCs and SVOCs in the presence of high arsenic concentrations, and (2) the removal of arsenic from the groundwater through a biological process using a unique bioreactor system.

METHODS AND MATERIALS

Groundwater Sampling and Characteristics

Groundwater from the Site was collected from several monitoring wells and placed into plastic 55-gallon drums for transport to the laboratory. Groundwater sampling was designed to represent the mixture of groundwater that a full-scale bioreactor would receive under field conditions. The pH of the groundwater was 8.3 and had a total chemical oxygen demand of 650 mg/L. Specific organic chemicals detected in the composite groundwater sample were: 2-hexanone (3.2 mg/L), acetone (7.2 mg/L), methyl ethyl ketone (1.3 mg/L), toluene (4.0 mg/L), total xylenes (0.4 mg/L), benzoate (0.5 mg/L), and naphthalene (0.1 mg/L).

Biological Reactor Pilot Plant (Bioreactor)

The bioreactor used in this study has been described previously in an investigation of the biotreatment of chlorinated VOCs (Thomas & Leong 1991). The bioreactor was a continuous-flow, four-cell, anaerobic/aerobic, submerged, fixed-film bioreactor manufactured by Tri-Bio, Inc. of Allentown, Pennsylvania (see Figure 1). Each of the four reactor cells had a capacity of 5.1 L, providing a total reactor volume of 20.4 L. All of the reactor cells contained rigid polyvinyl chloride medium with a high surface area-to-volume ratio, resulting in a cell liquid volume of about 4 L.

For this investigation, the bioreactor was configured to operate in both the anaerobic and aerobic modes. The treatment process was designed so that the first two cells (Cells 1 and 2) functioned under anaerobic conditions and the last two cells (Cells 3 and 4) operated under aerobic conditions. Cells 1 and 2 of the bioreactor were bypassed during a portion of the study in which only the aerobic mode was evaluated. The bioreactor was operated at ambient temperatures (21 to 24°C).

Anaerobic Cells. The first two cells of the bioreactor were inoculated with anaerobic digester sludge and then sealed from the atmosphere. High-fructose corn syrup was added into Cell 1 to promote a highly reducing environment. The bioreactor was acclimated for a period of 3 weeks before feeding groundwater into the system. The anaerobic condition was evaluated by measuring the oxidation-reduction potential (ORP) in Cell 1, and by qualitatively assessing the presence of methane gas. Cell 2 received the effluent from Cell 1 and also was maintained in an anaerobic condition.

The anaerobic conditions in Cells 1 and 2 were established to create a highly reducing environment that could promote the removal of arsenic as elemental arsenic and arsenide. Some arsenic removal was also expected to occur through its sorption by the microbial biomass.

Aerobic Cells. The last two cells of the bioreactor, Cells 3 and 4, were sustained in an aerobic condition (dissolved oxygen [DO] greater than 2 mg/L) by leaving the top of the reactor cells open to the atmosphere and pumping air into each cell. An aerobic condition was maintained to (1) catalyze the biodegradation of hydrocarbons flowing from the anaerobic cells, and (2) oxidize arsenic to arsenate, which is the form most amenable to precipitation with iron (Ghosh 1986, Korte & Fernando 1991).

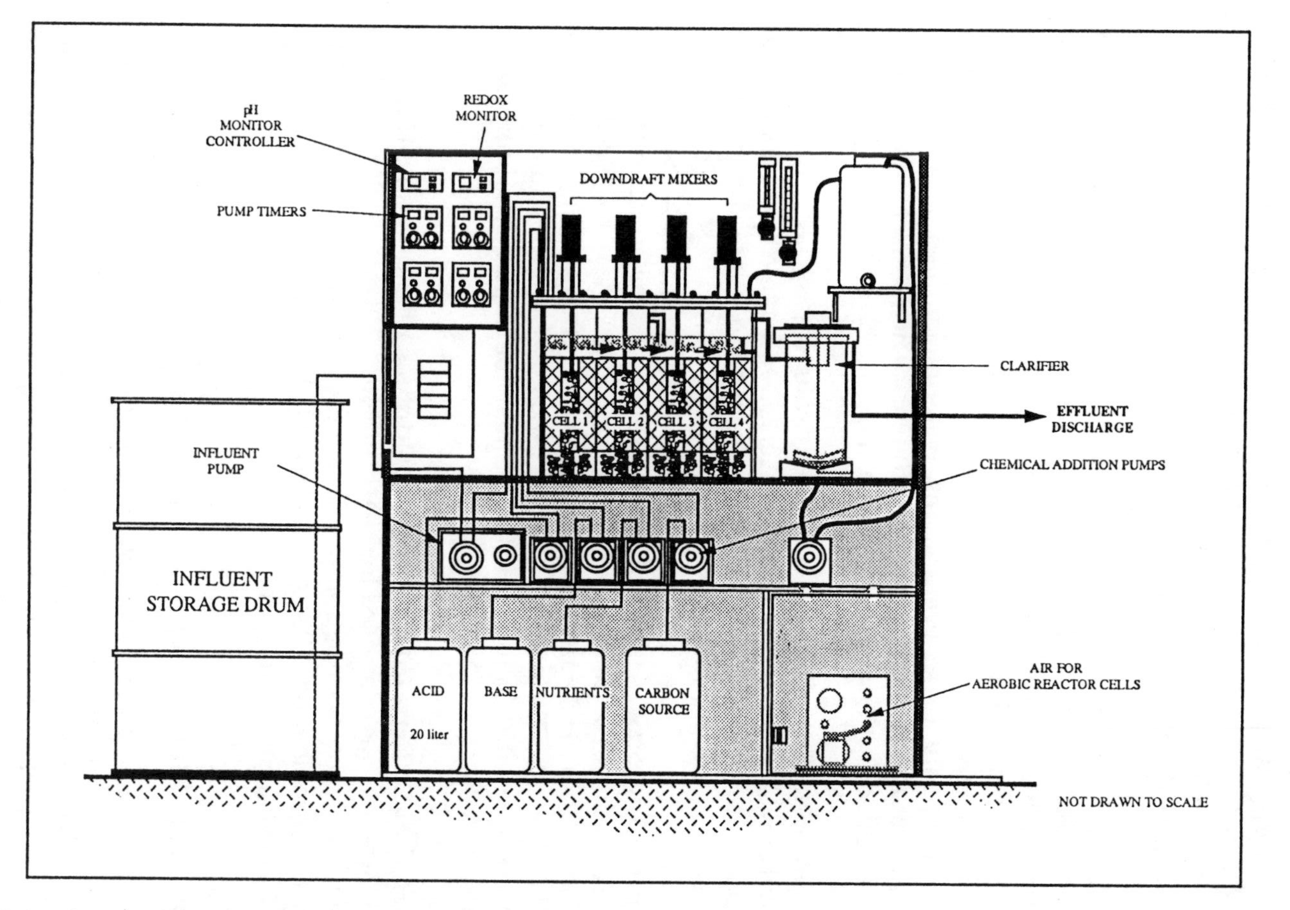

FIGURE 1. Anaerobic/aerobic submerged fixed-film bioreactor pilot plant.

Removal of arsenic in the aerobic cells was not anticipated during standard operating conditions because the oxyanions of arsenic, arsenite and arsenate, are extremely soluble in water. Therefore, ferric iron chloride ($FeCl_3$) was added to either Cell 3 or Cell 4 to provide ferric hydroxide surfaces for arsenic adsorption/precipitation reactions to occur.

Nutrients and Supplemental Carbon Sources. The nutrient solution was prepared by dissolving concentrated phosphoric acid and ammonium hydroxide into tap water at concentrations previously determined in our laboratory to be optimum levels for microbial growth in the bioreactor. Nutrients were added in amounts so that ammonia-nitrogen (N) and ortho-phosphate (P) concentrations were maintained between 2 to 10 mg/L in the filtered bioreactor effluent. The supplemental carbon source added to the bioreactor was a high-fructose corn syrup.

Spike Solution. The influent groundwater was spiked with acetone, methyl ethyl ketone (MEK), toluene, xylene, 2-hexanone, and naphthalene to compensate for the loss of VOCs during collection, shipment, and storage of the groundwater.

pH. Control of pH in the first anaerobic cell was critical because of the production of organic acids from acetogenesis. The pH was controlled to near-neutral pH conditions in Cell 1, whereas there was no pH control in Cells 2 and 3. The pH in Cell 4 was lowered from pH 8 (uncontrolled pH) to pH 5 with 0.2 mol/L HCl to evaluate the impact of pH on arsenic removal by $FeCl_3$.

Hydraulic Retention Time. The hydraulic retention time (HRT) in the bioreactor was controlled to range from 46 to 24 hours in the combined anaerobic/aerobic configuration. When the bioreactor was operated in only the aerobic mode (i.e., anaerobic Cells 1 and 2 were bypassed), the HRT was controlled to range from 36 to 8 hours.

Bioreactor Monitoring

The influent groundwater was sampled after the addition of nutrients, feed, and spike solutions, which diluted the influent groundwater by about 30%. The effluent was collected from Cell 4. The pH of each cell was measured immediately following sample collection. The samples were filtered (Whatman #40) and analyzed for: chemical oxygen demand (COD), using a Hatch COD kit; soluble nitrogen (N), using a Chemetrics colorimetric assay kit (Model No. K-1510); and soluble ortho-phosphate (P), using a Chemetrics colorimetric assay kit (Model No. PO-10).

Other parameters monitored in specific cells included ORP in Cell 1, dissolved oxygen (DO) in Cells 3 and 4, and total hydraulic flow from the Cell 4 effluent point. The ORP was measured in Cell 1 using a probe, which was inserted through the top of Cell 1 and was connected to a panel-mounted ORP monitor. DO was measured in the Cell 3 and Cell 4 liquid using a YSI DO meter and probe. The hydraulic flow was determined by diverting the effluent flow from the clarifier and measuring the volume collected during a measured period of time. The total suspended solids (TSS) of the effluent were determined using American Public Health Association (APHA) Standard Methods, 15th edition, Method 209C.

The concentrations of VOCs and SVOC in the bioreactor liquids were evaluated by EPA Methods 8240 and 8270, respectively. Emissions of VOCs and SVOCs from Cells 3 and 4 (vents in Cells 1 and 2 were sealed) were collected on carbon tubes over a period of 24 hours, extracted with methylene chloride and analyzed as above. Total arsenic in the filtered (Whatman #40) bioreactor liquids was measured using EPA Method 7060 (Graphite Furnace). Arsenic in the headspace gases from Cells 1 and 2, and in the emissions from Cell 3 was collected in an alkaline peroxide trap consisting of 4 parts 0.5 mol/L NaOH and 1 part 70% hydrogen peroxide and analyzed as above. Biomass from the bioreactor cells was collected after completion of the study, oven-dried at 100°C, acid-digested by EPA Method 3050, and analyzed for arsenic as above.

RESULTS AND DISCUSSION

Bioreactor Monitoring

The results of monitoring for ORP, pH, COD, P, N, DO, and TSS in the bioreactor cells during the course of the study are summarized in Table 1. The redox condition of the anaerobic cell was in the range in which arsenic can be transformed to elemental arsenic and arsenide, which are less soluble than the more oxidized forms (Alexander 1977; Dragun 1988; O'Niell 1990). The decreases in COD, N, and P measured across the bioreactor indicate microbial activity.

VOC and SVOC Removal

The total concentrations of VOCs and SVOCs in the influent (adjusted for dilution by supplementary solutions) and effluent during operation

TABLE 1. Summary of monitoring parameters in the bioreactor.

Parameters	Cell 1	Cell 2	Cell 3	Cell 4
ORP (mV)	−355 ± 98	NA	NA	NA
pH	5.3 to 7.4	6.4 ± 0.7	7.5 ±0.5	5.1 to 8.1
COD	496 ± 93	392 ± 83	160 ± 97	70 ±40
NH_3-N (mg/L)	33 ± 17	33 ±19	14 ± 9	12 ± 5
PO_4-P (mg/L)	7.2 ± 2.5	8.3 ± 3.4	1.6 ± 1.9	0.4 ± 0.1
DO (mg/L)	NA	NA	5.1 ± 1.4	5.6 ± 1.3
TSS (mg/L)	NA	NA	NA	35 ± 22

Data are mean values ± 1 standard deviation, except for the range of pH values in Cells 1 and 4.

of the bioreactor in the combined anaerobic/aerobic mode and the aerobic-only mode are displayed in Figure 2. The results show that the bioreactor in the anaerobic/aerobic configuration has the capability of removing VOCs and SVOCs to below detectable levels (0.001 to 0.02 mg/L) for all compounds detected by EPA Methods 8240 and 8270. Operating at an HRT of about 30 hours, the bioreactor produced consistently nondetectable concentrations of VOCs and SVOCs in the effluent. When the HRT was decreased to 24 h, detectable concentrations of 2-hexanone, acetone, and MEK appeared in the effluent. This breakthrough of organics, however, may have been a result of the lowering of pH in Cell 4 (lowered to pH 5 to affect arsenic adsorption), which began about 1 week before the detection of organics in the effluent. After adjusting the Cell 4 pH to above 6, the concentrations of VOCs and SVOCs in the effluent were again observed to be below their limits of detection.

The bioreactor was also operated with only Cells 3 and 4 in operation (aerobic-only mode). The concentrations of VOCs and SVOCs in the influent (corrected for dilution by supplemental solutions) and effluent during this type of configuration also are illustrated in Figure 2. The results show that the bioreactor, in this configuration, was able to remove VOCs and SVOCs to nondetectable concentrations. Emissions of VOCs and SVOCs from Cells 3 and 4 were not detected in the carbon tubes attached to the vent ports, indicating that biodegradation was probably the primary mechanism for removing these organic compounds in this bioreactor.

Arsenic Removal

The concentrations of arsenic in the influent and effluent during operation of the bioreactor in the combined anaerobic/aerobic mode are

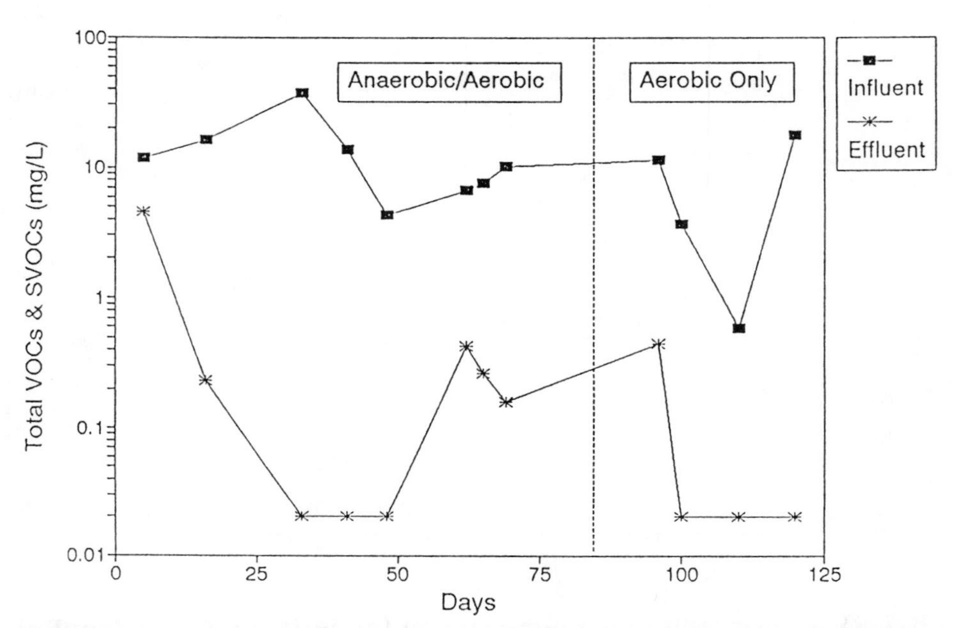

FIGURE 2. Concentrations of total VOCs and SVOCs in the influent and effluent of the bioreactor configured in the anaerobic/aerobic and aerobic-only modes.

illustrated in Figure 3. The influent concentrations of arsenic (corrected for dilution by supplemental solutions) ranged from 45 to 70 mg/L, whereas the effluent levels of arsenic ranged from 0.2 to 25 mg/L. The profile of arsenic concentrations in each of the bioreactor cells is illustrated in Figure 4. Most of the arsenic removal was observed to occur in Cell 1 where arsenic concentrations were reduced by more than 70%. Arsenic removal in the anaerobic cells may have been due to a variety of factors, including reduction to elemental arsenic, coprecipitation with sulfides, and sorption by microbial biomass.

After the addition of $FeCl_3$ to Cell 3 (resulting in an Fe concentration of 50 to 100 mg/L), the removal of arsenic was observed to increase to almost 90% as the effluent arsenic concentrations declined to about 2 mg/L. Because iron precipitation with arsenic is known to be greater with arsenic in the most oxidized form (arsenate), additions of $FeCl_3$ were then moved to the second aerobic cell (Cell 4). The results presented in Figure 3 show that this addition of $FeCl_3$ decreased the concentrations of arsenic in the effluent to about 0.20 mg/L. The greater efficacy of arsenic removal during the addition of $FeCl_3$ to Cell 4, rather than to Cell 3, indicates that the more oxidized form of arsenic, arsenate, is more

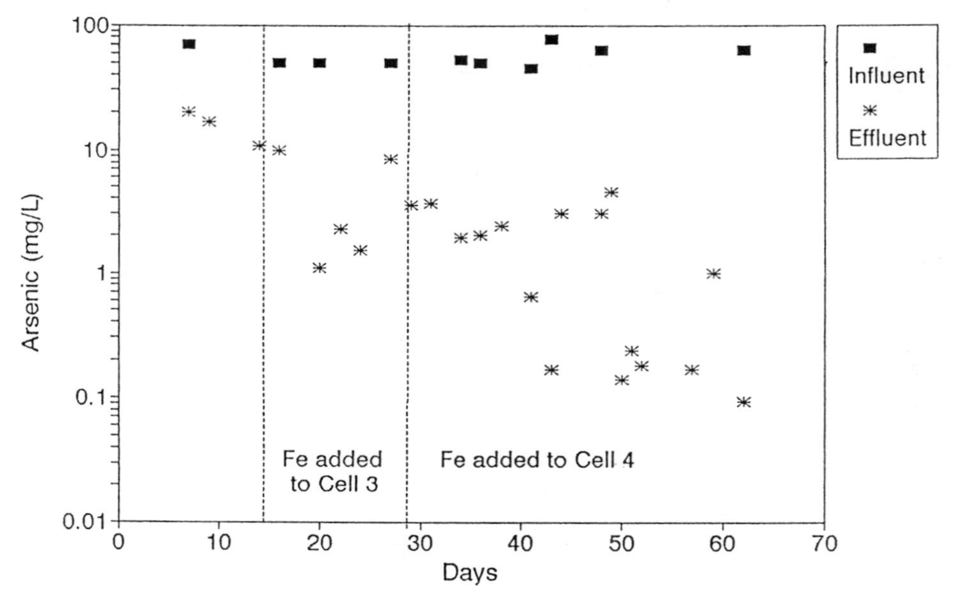

FIGURE 3. Concentrations of arsenic in the influent and effluent of the bioreactor after additions of $FeCl_3$ to Cell 3 and then to Cell 4.

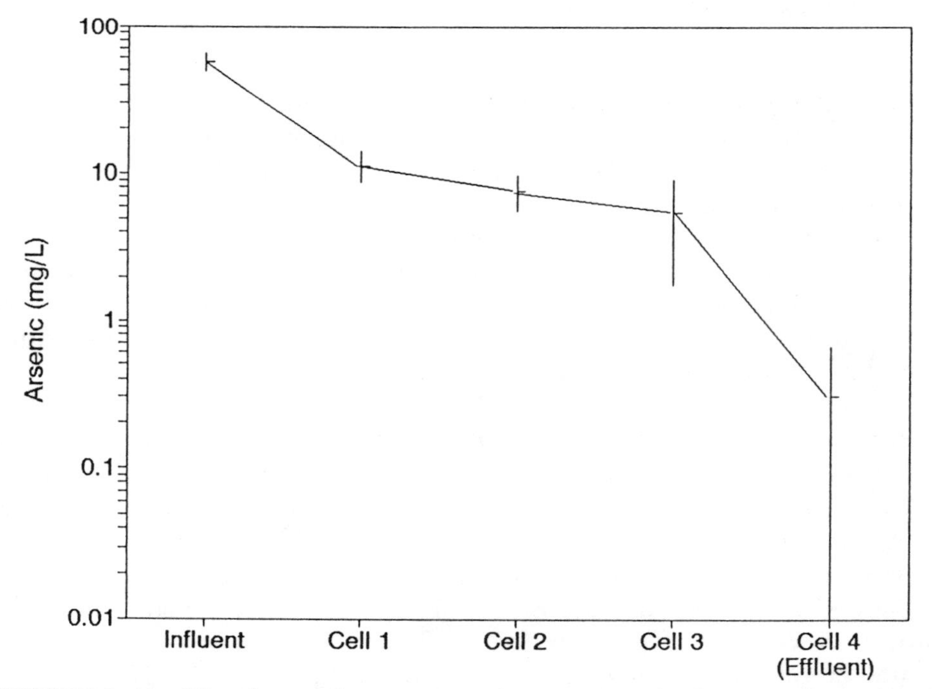

FIGURE 4. Profile of arsenic concentrations across the bioreactor. Error bars represent 95% confidence limits (n = 6-17).

amenable to removal by Fe than the less oxidized form, arsenite. This supposition is comparable to other investigations of arsenic treatment by iron salts (Ghosh 1986). Lowering pH from 6.9 to 5.3 in Cell 4 was not observed to have any significant impact on the concentration of arsenic in the effluent. During this period of optimum arsenic removal, concentrations of VOCs and SVOCs in the effluent remained below their limits of detection.

The amount of arsenic volatilized from Cells 1, 2, and 3, as a percentage of the total mass loading of arsenic into the bioreactor, was determined to equal 0.005%, 0.057%, and 0.018%, respectively. These results indicate that some volatile forms of arsenic are formed within the bioreactor, possibly arsines in the anaerobic cells and methyl arsenic compounds in the aerobic portion (Alexander 1977). However, the low levels of arsenic volatilized from the bioreactor reveal that these forms of arsenic do not contribute significantly to the overall removal of arsenic from the groundwater.

Upon completion of the study, the biomass solids accumulated in the bioreactor cells were collected and analyzed for arsenic. The concentrations of arsenic measured in the oven-dried solids from Cells 1, 2, 3, and 4 were found to be 32,000 mg/kg, 25,000 mg/kg, 6,900 mg/kg, and 1,900 mg/kg, respectively. The large concentrations of arsenic in the biomass indicate that substantial amounts of arsenic were associated with the microbial biomass and that microorganisms can survive and effectively degrade VOCs and SVOCs in the presence of extremely high levels of arsenic.

CONCLUSIONS

The results of the biotreatability study showed that VOCs and SVOCs in the groundwater can be treated to below laboratory detection limits in the presence of high arsenic concentrations. Treatment of VOCs and SVOCs using this type of bioreactor demonstrated effectiveness in both a combined anaerobic/aerobic configuration and in a strictly aerobic configuration. The removal of arsenic from groundwater in the bioreactor configured in the anaerobic/aerobic mode was greater than 99%. More than 70% of the arsenic was removed in the anaerobic cells and effluent concentrations of arsenic were lowered to less than 0.2 mg/L through the addition of $FeCl_3$ in aerobic Cell 4. No significant amounts of arsenic were removed in the aerobic cells without the addition of $FeCl_3$. With additional optimization of $FeCl_3$ and pH in the aerobic cells and pH/redox conditions in the anaerobic cells, arsenic removal in this bioreactor may be enhanced further.

REFERENCES

Alexander, M. 1977. *Introduction to Soil Microbiology.* John Wiley & Sons, New York, NY.

Dragun, J. 1988. *The Soil Chemistry of Hazardous Materials.* Haz. Mater. Contr. Res. Inst., Silver Springs, MD.

Frost, R. R., and R. A. Griffin. 1977. "Effect of pH on adsorption of arsenic and selenium from landfill leachate by clay minerals." *Soil Sci. Am. J. 41*:53.

Ghosh, M. M. 1986. "Adsorption of inorganic arsenic and organoarsenical on hydrous oxides." In J. W. Patterson and R. Passino (Eds.), *Metals Speciation, Separation, and Recovery*, p. 499. Proc. Int. Symp. Metals Spec. Sep. Rec. Chicago, IL. Lewis Publishers, Chelsea, MI.

Korte, N. E., and Q. Fernando. 1991. "A review of arsenic (III) in groundwater." *Crit. Rev. Environ. Contr. 21*(1):1.

O'Niell, J. 1990. "Arsenic." In B. J. Alloway (Ed.), *Heavy Metals in Soils,* p. 83. Blackie, London, and John Wiley & Sons. New York, NY.

Thomas, J. M., and G. M. Leong. 1991. "The biological treatment of groundwater contaminated with chlorinated and nonchlorinated volatile organic compounds: A continuous-flow pilot-scale treatability study." Paper 91-19.1, 20 pp. Air Waste Manage. Assoc., Vancouver, BC.

RECYCLING OF COPPER FLOTATION TAILINGS AND BIOREMEDIATION OF COPPER-LADEN DUMP SITES

C. F. Gökcay and S. Önerci

ABSTRACT

Many mining wastes contain valuable metals, but recovering them by conventional mining technology often is uneconomical. Although metal contents are rather low in these wastes, their total productions are often high. At present, metal contamination of surface and groundwaters from such dumping sites is a serious cause for concern. Development of an economical and simple process to treat such residues is desirable for at least two reasons: valuable metals can be recovered from these waste materials, and environmental pollution can be decreased. This paper discusses the applicability of a bacterial leaching technique for the copper ore flotation tailings produced at the Ergani-Maden mining and copper processing plant in Turkey. At present, 400,000 metric tons of finely ground (-65 mesh) flotation tailings with 0.4% copper contents are being dumped annually to a nearby dumping site, resulting in pollution of the surface and groundwaters in that region. Flotation tailings, with their 3% sulfur and 30% iron contents, make ideal substrates for thiobacilli. During lab-scale leach experiments using *Thiobacillus ferrooxidans*, the effects of several parameters on copper-leaching efficiency were examined.

INTRODUCTION

Leaching of metals by simply percolating water through columns or heaps of finely ground ores has long been practiced by hydrometallurgists to enrich the ores and ultimately to recover these metals. Although the process dates back to the 1600s or earlier, the presence of bacteria in the leaching fluids was not confirmed until 1962 (Razzel & Trussell 1963). Since then, efforts have been made to enhance bacterial capability for more efficient leaching.

Overt exploitation of metal deposits during the last century has led to the accumulation of large heaps of metal-bearing wastes that threaten underground and/or surface waters that are recipients of drainage from such deposits. In the past, waste volumes were small, land was plentiful, and the environment was not considered critical. Today reclamation of land laden with these metallic wastes and prevention of water pollution caused by them is a serious issue. Biohydrometallurgical approaches may offer the most promising solution. Metals leached from waste heaps by such simple and low-cost processes may then be recovered from liquors and recycled.

At the Ergani-Maden copper mining site, in Turkey, the copper ore is extracted from open pits. The ore is crushed, finely ground, and conveyed to flotation cells where the copper portion is separated from the gangue material. The residues from flotation, i.e., the tailings, still contain appreciable amounts of copper and typically are discarded on land in large heaps. Fresh tailings are not inert and tend to leach naturally, contaminating the groundwater and a nearby stream. High concentrations of copper and extremely low pH are common in this stream. The plant annually produces about 400,000 tons of flotation tailings, and the disposal site already contains 4 to 5 million tons of this material. A chemical analysis of a flotation tailings sample is presented in Table 1.

Among the microorganisms involved in the industrial production of various metals, *Thiobacillus ferrooxidans* is the most widely used. *T. ferrooxidans* reportedly is able to effect the dissolution of more than 90% of

TABLE 1. Chemical analysis of Ergani-Maden collective flotation tailings (particle size: -65 mesh).

Element/Compound	% by Weight
Fe	30.5
S	3.0
SiO_2	27.6
MgO	8.0
Cu	0.4
Al_2O_3	13.0

copper from common sulfidic ores such as chalcopyrite and chalcocite. The bacterium is a mesophilic, Gram-negative, chemolithoautotroph. It is capable of fixing carbon dioxide through the Calvin cycle (Gale & Beck 1967) while oxidizing reduced forms of sulfur, iron, and metallic sulfides in the presence of oxygen.

The bacterium produces sulfuric acid as a result of biological oxidation. It can withstand harsh environmental conditions such as high metal concentrations and low pH, to pH 1, although pH 2.5 to 4.5 is optimum (Ehrlich & Fox 1967). Above pH 3.0, ferric iron precipitates. The reduced leaching rates observed at or above pH 3.0 are thought to be due to either binding of the sulfide surface by the precipitate or depletion of the ferrous iron in solution to effect rapid oxidation of the sulfidic mineral. Duncan et al. (1966) have shown that the organism can oxidize sulfur at pH as high as 5.0, but sulfides never above pH 4.0.

Although *T. ferrooxidans* is the most common bacterium involved in the bacterial leaching of metals, it is by no means the only microorganism with that capacity. Among other organisms, *T. thiooxidans*, *Leptospirillum ferrooxidans*, and a number of thermophiles of the genus *Sulfolobus* deserve mention. The latter are mixotrophs relying on organic carbon for biosynthesis and inorganics, such as sulfur, for energy.

The exact mechanism of bacterial leaching is poorly understood. Two have been postulated: a direct-contact mechanism and an indirect-contact mechanism. The direct-contact mechanism assumes physical contact between the bacteria and sulfide minerals under aerobic conditions (Silver 1978). The direct-contact mechanism was favored by Beck and Brown (1968), who showed that both iron and sulfur are attacked by bacteria in pyrite oxidation. In support of this view, Torma (1972) reported that the bacterium is able to oxidize iron-free, analytically pure cobalt, nickel, and zinc sulfides. Berry and Murr (1978) published electron micrographs of *T. ferrooxidans* attached to chalcopyrite crystals.

In the indirect mechanism, the role of bacteria is considered not to attack pyrite directly but to catalyze aerobic oxidation of Fe^{+2} in solution to Fe^{+3}. The Fe^{+3} ion in solution then oxidizes pyrite to Fe^{+2}, and additional acidity is produced due to the protons produced in this reaction (Brierley 1982).

Factors affecting the rate and efficiency of metal leaching by thiobacilli are temperature, particle size, pulp density, initial pH, inoculum size and age, nutrients, mineralogy of the ore, and the presence and concentrations of various metals.

Extensive culturing of *T. ferrooxidans* was first achieved by Silverman and Lundgren (1959) on a completely mineral growth medium containing 9 g/L of iron. This medium, called "9K" medium, contains the usual elements needed by bacteria, such as nitrogen and phosphorus. Similar media have since been devised as variants of 9K. One with the acronym LOPOSO (because it contains low phosphate and low sulfate) was introduced by Hoffmann et al. (1981) to overcome the uncertainty in turbidimetric analysis that occurred with high sulfate concentrations. The chemical compositions of both media are given in Table 2. At least two strains of *T. ferrooxidans* reportedly are able to fix molecular nitrogen (Mackintosh 1976).

Most heterogeneous reactions proceed at an initial rate proportional to the surface area of the phase boundary. Theoretically, the rate will continue to rise with increasing surface area until a nonsurface reaction controls the rate. The total mineral surface area available to the surface reactions may be increased either by decreasing the particle size or by increasing the solids concentration. Both the rate and extent of leaching have been increased for up to a certain solids concentrations (Duncan et al. 1966 and Torma et al. 1972). It is postulated by these workers that slower rates observed at higher pulp concentrations were due to enhanced interference of particles with the transfer of oxygen and carbon dioxide to the organism. Another explanation for this paradoxical phenomenon came from Kelly and Jones (1978), who suggested that the lower rates may be a manifestation of substrate and/or product inhibition.

It is generally agreed that the oxidation of mineral sulfides to sulfates and the dissolution of copper by bacteria are both first-order phenomena, yielding a straight-line relationship when the logarithm of concentration is plotted versus time. Chakraborti and Murr (1980) estimated the order of the bacterial leaching reactions to be 0.92.

In spite of the relative resistance of chalcopyrite to biological leaching, earlier reports indicate that bioleaching of this mineral in tailings is possible (Ebner 1978, Groudev et al. 1980). Groudev et al. (1980) obtained

TABLE 2. Growth media for *Thiobacillus ferrooxidans*.

Components g/L	9K	LOPOSO
$(NH_4)_2SO_4$	3.0	—
K_2HPO_4	0.5	0.1
KCl	0.1	0.1
$MgSO_4$, $7H_2O$	0.5	0.85
$Ca(NO_3)_2$	0.01	0.008
$FeSO_4$, $7H_2O$	44.2	—

more than 65% copper extraction from tailings under tightly controlled experimental conditions.

The commonly employed bacteriological methods do not lend themselves to the study of iron-oxidizing bacteria. For example, it is not possible to grow iron bacteria on agar plates, nor is it possible to grow them in classical rich medium, such as nutrient broth or glucose broth. Frequently, studies have been based on following and evaluating net changes that occur in a controlled environment, such as in a shake flask or in a percolator, caused by the bacteria present in that system. This approach has considerable practical value, for it generally provides the rates of reactions needed for quantitative evaluation.

MATERIALS AND METHODS

The shake flask method was chosen in this study to test whether the Ergani-Maden collective copper tailings are amenable to extraction by bacteria. This method was chosen partly because numerous experiments could be carried out simultaneously in a relatively simple set up, and partly because it provides the necessary kinetic data to design lagoons that are necessary for on-site treatment of tailings, as well as providing primary data for the in situ treatment of the dump sites at Ergani. The conditions giving maximum rates and extents of copper extraction were sought by designing six sets of batch experiments.

Factors tested in these experiments were the effectiveness of the two different *Thiobacillus ferrooxidans* strains and the effects of medium composition, initial pH, inoculum age, temperature, particle size, and percent solids concentration (pulp density). Factors were tested at three to four levels, as presented in the Results and Discussion section. For each factor, a total of six samples were drawn from supernatants of flasks at 3- to 4-day intervals, until the 22nd day was reached. These samples were analyzed for soluble copper. The copper dissolution rate constants were evaluated by plotting logarithms of measured copper concentrations versus time and fitting a straight line through these points by least squares. The slope of this line is calculated by the least squares algorithm, thereby giving the rate constant.

The chemical analysis of the original tailing sample, of -65 mesh (-210 µm) particle size, is given in Table 1. The four other particle sizes (D_p) — 210 µm > D_p > 105 µm, 105 µm > D_p > 53 µm, 53 µm > D_p > 37 µm, and D_p < 37 µm — were prepared by further grinding the original sample and sieving on a shaking screen. Then they were washed with dilute acid (pH 2), rinsed with distilled water, and dried to remove any residual lime (CaO) left from the flotation process.

The strain of *Thiobacillus ferrooxidans* (strain no. OSU 380), designated as "Ohio strain," was provided by P. R. Dugan, and the "Mexico strain" was provided by C. L. Brierley. Stock culturing of these strains was undertaken on LOPOSO liquid medium containing finely ground tailings (-300 mesh). The composition of the LOPOSO medium is given in Table 2. During the leaching experiments, if not otherwise stated, 100 mL liquid LOPOSO medium, adjusted to pH 3.0 and containing 5% original flotation-size (-65 mesh) tailings, were placed into 500-mL conical flasks. Flasks were then inoculated with 2 mL of 15-day-old stock culture and incubated at 30°C, shaking, for 22 days. A period of 22 days was chosen arbitrarily, based on the lead experiments, where the rate of copper extraction leveled off after the 20th day. In some of the later experiments, copper extraction continued at high rates even after 22 days, but no attempts were made to follow this to completion. A color change in the test flasks, from yellow to red, and a drop in pH signified bacterial activity and metal solubilization. The initial pH and color in the uninoculated control flasks remained the same throughout the experiments. The control flasks also contained 1% mercury chloride to stop any bacterial activity.

At the end of the experiments, the tailing samples were filtered, washed with distilled water, and dried. The difference between the initial and residual copper content at the end of experiment gave the amount of copper extracted. Solid samples were digested with concentrated HCl + HNO_3 (1 + 3 by vol.) acids to determine their copper content by atomic absorption spectroscopy (AAS). A Perkin Elmer AAS with ultraviolet (UV) background correction was used for the AAS analysis. During incubation, samples were drawn from flasks at regular intervals and centrifuged. The amount of copper in solution was determined by directly aspirating acidified supernatant samples (pH 2 with HNO_3) into the air/acetylene flame. A material balance was conducted at the end of each experiment for verification. The pH of the medium was measured every 1 to 2 days by dipping a pH electrode into the flasks.

RESULTS AND DISCUSSION

Effect of *Thiobacillus* Strain and Growth Medium

To determine the effect of the *Thiobacillus* strains and medium composition on copper extraction efficiency, the Ohio and Mexico strains were tested in 5% (wt/vol) tailing slurries (referred to as pulp density), at 30°C, using two different growth media. The one-way analysis of the copper extraction results that are shown in Table 3 indicated with

TABLE 3. Effect of different *Thiobacillus* strains and growth media on bioleaching of copper from tailings.

			Rate constants, h^{-1}		
Inoculum	Medium	$HgCl_2$	$Cu^{(a)}$	$S^{=\,(b)}$	% Cu extraction
Ohio	9K	−	0.003	0.00218	55
—	9K	+	0.0018	0.00074	18
Mexico	9K	−	0.0029	0.0019	53
—	9K	+	0.00124	0.0005	14
Ohio	LOPOSO	−	0.00345	0.004	70
—	LOPOSO	+	0.00124	0.00108	17
Ohio	D.WTR	−	0.0028	0.00156	50
—	D.WTR	+	0.00093	0.00093	12.5

Initial pH = 2.5; 30°C; -65 mesh D_p; 2 mL 15-day-old stock culture; 5% pulp density.
(a) Rate constant for copper solubilization.
(b) Rate constant for sulfide oxidation.

95% confidence that treatment corresponding to Ohio strain and the LOPOSO medium was significant amongst the others. However, statistically justifiable distinction between responses of the remaining factors could not be made. Thus the LOPOSO medium and Ohio strain were used in the subsequent experiments.

Effect of Initial pH

The initial pH is considered important in leaching experiments, and the natural drop of pH as leaching progresses is known to be preferable to keeping it constant by alkali addition. The most suitable initial pH was sought in these experiments by incubating four test flasks in parallel with corresponding sterile controls. The pH was adjusted prior to initiation of each experiment, but slight pH adjustments were still needed in the early days using 0.1 N sulfuric acid, to overcome the neutralizing effect of lime carried over from the previous flotation process. An initial pH of 3.0 seemed to give higher copper extraction rate and extent as shown in Table 4. However, this effect could not be substantiated by statistics as one-way analysis of the paired data or variance analysis of the pooled data did not indicate significance of any of the treatments over the others with 95% confidence. Nevertheless further experiments were carried out at an initial pH of 3.0.

TABLE 4. Effect of initial pH on the bioleaching of copper from tailings.

Initial pH	Rate constant, h^{-1} Cu[a]	% Cu extraction
2.0	0.00092	66
2.5	0.00143	69
3.0	0.003	74

Inoculum = Ohio; 30°C; -65 mesh D_p; 2 mL 15-day-old stock culture; 5% pulp density; LOPOSO medium.
(a) Rate constant for copper solubilization.

Effect of Inoculum Age

Each of the test flasks received 2 mL inoculum from supernatants of decanted 5-, 10-, 15-, and 20-day-old cultures respectively, grown on 5% tailings in LOPOSO medium. The results of these experiments are shown in Table 5. The one-way, pair-wise analysis of results indicated with 95% confidence that the effects of treatments on copper dissolution are significant, with the exception of the comparison between 10-day and 20-day-old cultures.

The effect of inoculum age appears to be related to the concentration of cells inoculated in to the test flasks. For example, the cell counts with the most probable number (MPN) technique using LOPOSO medium revealed that the cell concentration in suspension, which was 10^4 cells/mL in 5-day-old cultures, increased to 10^7 cells/mL on the 10th day. The cell count was exceeding 10^8 cells/mL in 15-day-old cultures. As seen in Table 5, that growth in this system ceased after the 15th day and a stationary phase with concomitant dissociation of cells was experienced, as the rate and extent of copper solubilization with 20-day-old inoculum was equal to that for 10-day-old inoculum, which was significantly lower than that for the 15-day-old inoculum. The data also suggest that mineral surfaces are not saturated with bacteria for up to, at least, 2×10^6 cells/mL, as understood from 15-day-old culture data. A higher inoculum size from 15-day-old culture might have provided faster rates but was not tested in this study.

Effect of Temperature

The effect of temperature on the bioleaching of tailings was tested by incubating test flasks at three temperatures: 25°C, 30°C, and 35°C.

Table 6 shows the results of these experiments. The one-way, pair-wise statistical analysis of results indicated with 95% confidence that the effect of temperature on copper dissolution is significant within the interval between 25° and 30°C. No distinction could be made between 30°C and 35°C, indicating that a plateau is reached after 30°C.

The activation energy and temperature quotient for copper solubilization were calculated as –13.2 kcal/mole and 2.03 respectively, assuming a straight-line relationship exists between 25° and 30°C when log values of rate constants were plotted versus the reciprocal of temperatures in °K.

Effect of Particle Size

Four different particle-size fractions with a fixed pulp density (5%) were tested for copper extraction. Each test flask received 2 mL inoculum from the 15-day-old culture. Flasks contained tailing samples of -65 mesh size in LOPOSO medium. The one-way, pair-wise, statistical analysis of the results that are shown in Table 7 indicated with 95% confidence that the effects of the tested particle sizes on copper dissolution are significant. Comparison of the data in Table 7 with those in Table 6 indicates that rate constant and copper extractions obtained for the -270 +400 mesh fraction, as shown in Table 7, were almost identical to those obtained with tailing-size particles at 30°C, in Table 6.

In practice, the optimum particle size has to be determined for each kind of substrate to be leached (Torma 1977). This size is dictated by the economics of the grinding process considering gains in leach rate versus increased cost of grinding to obtain smaller particle sizes. At the high cost of grinding, it is debatable whether regrinding of tailings

TABLE 5. Effect of inoculum age on the bioleaching of copper from tailings.

Inoculum age (days)	Rate constant, h^{-1} $Cu^{(a)}$	% Cu extraction
5	0.00246	34
10	0.00263	56
15	0.00293	73
20	0.0025	54

Inoculum = Ohio; 30°C; -65 mesh D_p; 5% pulp density; LOPOSO medium; pH_i = 3.0.

(a) Rate constant for copper solubilization.

TABLE 6. Effect of temperature on the bioleaching of copper from tailings.

Temperature (°C)	Rate constant, h^{-1} $Cu^{(a)}$	% Cu extraction
25	0.0015	32
30	0.00293	73
35	0.00305	75

Inoculum = Ohio; -65 mesh Dp; 2 mL 15-day-old stock culture; 5% pulp density; LOPOSO medium; pH_i = 3.0.
(a)Rate constant for copper solubilization.

to -400 mesh size is at all worth the effort, particularly when a mere 11% increase in copper extraction could be obtained by doing so. The need for sieving original tailings to obtain high efficiencies was not indicated in these experiments.

Effect of Pulp Density

The effect of pulp density on the microbial leaching of metals has been indicated by various researchers. To test its effectiveness in this case, four test flasks were prepared with 5%, 10%, 15%, and 20% pulp density. The results are shown in Table 8. The statistical, one-way ANOVA analysis of these results indicated a low significance level (i.e., 80% confidence) between treatments. However, pair-wise analysis of 15 and 20% pulp density indicated treatment effect with 90% confidence.

TABLE 7. Effect of particle size on the bioleaching of copper from tailings.

Particle size (mesh)	Rate constant, h^{-1} $Cu^{(a)}$	% Cu extraction
-65 +150	0.002	30
-150 +270	0.00277	50
-270 +400	0.00312	74
-400	0.00342	86

Inoculum = Ohio; 30°C; -65 mesh D_p; 2 mL 15-day-old stock culture; 5% pulp density; LOPOSO medium; pH_i= 3.0.
(a) Rate constant for copper solubilization.

TABLE 8. Effect of pulp density on the bioleaching of copper from tailings.

Pulp density (%)	Rate constant, h^{-1} $Cu^{(a)}$	% Cu extraction
5	0.00293	73
10	0.0048	76
15	0.00554	78
20	0.00342	68

Inoculum = Ohio; 30°C; -65 mesh D_p; 2 mL 15-day-old stock culture; LOPOSO medium; pH_i= 3.0.
(a) Rate constant for copper solubilization.

Of the four different pulp densities tested, 15% appeared to support the highest rate and extent of copper solubilization.

The results discussed until this point indicate that a copper extraction efficiency of 78% can be achieved with flotation tailings in 22 days, at 30°C, 15% pulp density, and with unground (-65 mesh) flotation-size samples. The low phosphate and low-sulfur LOPOSO medium, as well as 2% inoculum from a 15-day old culture of *Thiobacillus ferrooxidans* (>10^8 cells) is required for this outcome. A further increase to 86% or over in extraction efficiency is likely when tailings are reground to -400 mesh and pulp density is increased to 15%.

These findings are comparable with those of Groudev et al. (1980), who reported a copper extraction efficiency of 65 to 70% in 12 days at 35°C with -400 mesh tailings containing 0.1% copper. In our case, about 63% copper extraction is calculated for the same time period with -400 mesh tailings at 30°C. This figure could go up if experiments were carried out at 15% pulp density.

A two-fold benefit is envisaged in bioremediation of copper-laden sites in this example. First, removal of copper from flotation tailings is expected to cut considerably the contamination of surface and underground waters receiving drainage from this site. A total of 1600 tons of copper would normally be expected to find its way annually to water courses in the region from this site, should no action be taken.

Conversely, bioremediation of this site by simple in situ leaching techniques or on-site by using ponds would cut the copper contamination of waters by at least 70 to 80%, and the final residue would be more or less inert. Moreover, revenue would be generated from recovering approximately 1,200 tons of solubilized copper from the rich liquor, and this in turn should significantly reduce bioremediation costs. Considering the copper available in ore entering the flotation process in this plant is merely 1.3% of the ore, and one-third of this amount is lost to the tailings, the revenue from bioremediation should be considerable.

One cause for concern is the by-product acid produced during the bioleaching process. However, acidity would be produced naturally in the tailing dump sites, anyway. Bioreclamation of existing tailings in dump sites by in situ leaching and/or on-site biotreatment of those that are always generated anew from the flotation process, i.e., by lagooning, would facilitate neutralization of the acidic drainage which would otherwise drain uncontrolled into the receiving waters.

CONCLUSION

The copper flotation tailings are amenable to biological leaching with appreciable copper solubilizing into the liquid phase. Bioleaching may be carried out on site in ponds, or in situ at the dump sites. The near-optimum conditions for extracting of copper from tailings in suspended culture systems are described in this paper and the optimum for in situ leaching should not be far from those observed in this study.

REFERENCES

Beck, J., and D. G. Brown. 1968. "Direct sulphide oxidation in the solubilization of sulphide ores by *Thiobacillus ferrooxidans*." *J. Bacteriol. 96*:1433-1434.

Berry, V. K., and L. E. Murr. 1978. "Direct observation of bacteria and quantitative studies of their catalytic role in the leaching of low-grade, copper-bearing wastes." In L. E. Murr, A. E. Torma, and J. A. Brierley (Eds.), *Metallurgical Applications of Bacterial Leaching and Related Microbiological Phenomena*, pp. 103-136. Academic Press, New York, NY.

Brierley, C. L. 1982. "Microbiological mining." *Scientific American*, August, pp. 42-51.

Chakraborti, N., and L. E. Murr. 1980. "Kinetics of leaching chalcopyrite-bearing waste rock with thermophilic and mesophilic bacteria." *Hydrometallurgy. 5*:337-354.

Duncan, D. W., C. C. Walden, and P. C. Trussell. 1966. "Biological leaching of mill products, Joint Meeting of the B.C. Section and Merritt Branch, C.I.M., Merritt, B.C. – October 1965." *Transactions 69*:329-333.

Ebner, H. E. 1978. "Metal recovery and environmental protection by bacterial leaching of inorganic waste materials." In L. E. Murr, A. E. Torma, and J. A. Brierley (Eds.), *Metallurgical Applications of Bacterial Leaching and Related Microbiological Phenomena*, pp. 195-207. Academic Press, New York, NY.

Ehrlich, H. L., and S. I. Fox. 1967. "Environmental effects on bacterial copper extraction from low-grade copper sulphide ores." *Biotechnol. Bioeng. 9*:471-485.

Gale, N. L., and J. L. Beck. 1967. "Evidence for the Calvin Cycle and Hexose Monophosphate Pathway in *Thiobacillus ferrooxidans*." *J. Bacteriol. 94*: 1052-1059.

Groudev, S., P. Genchev, and V. Groudeva. 1980. "Bacterial leaching of copper from flotation plant tailings." *Proc. 3rd Balkan Mineral Processing Conference*, pp. 195-199. Belgrade, Yugoslavia, April 23-26.

Hoffmann, M. R., B. B. Faust, F. A. Panda, H. H. Koo, and M. Tsuchiya. 1981. "Kinetics of the removal of iron pyrite from coal by microbial catalysis." *Appl. Environ. Microbiol. 42*: 259-271.

Kelly, D. P., and C. A. Jones. 1978. "Factors affecting metabolism and ferrous iron oxidation in suspensions and batch cultures of Thiobacillus ferrooxidans." In L. E. Murr, A. E. Torma, and J. A. Brierley (Eds.), *Metallurgical Applications of Bacterial Leaching and Related Microbiological Phenomena*, pp. 19-45. Academic Press, New York, NY.

Mackintosh, M. E. 1976. "Fixation of 15N2 by *Thiobacillus ferrooxidans*." *Proc. Soc. Gen. Microbiol. 4*:23-27.

Razzel, W. E., and P. C. Trussell. 1963. "Isolation and properties of an iron oxidizing *Thiobacillus*." *J. Bacteriol. 85*: 595-603.

Silver, M. 1978. "Metabolic mechanism of iron oxidizing *Thiobacillus*." In L. E. Murr, A. E. Torma, and J. A. Brierley (Eds.), *Metallurgical Applications of Bacterial Leaching and Related Microbiological Phenomena*, pp. 3-14. Academic Press, New York, NY.

Silverman, M. P., and D. G. Lundgren. 1959. "Studies on the chemoautotrophic iron bacterium *Ferrobacillus ferrooxidans*." *J. Bacteriol. 77*: 642-647.

Torma, A. E. 1972. "Microbiological Extraction of Cobalt and Nickel from Sulphide Ores and Concentrates." Canadian Patent Number 1,382,357.

Torma, A. E. 1977. "*Thiobacillus ferrooxidans* in hydrometallurgical processes." In T. Ghose, A. Fiechter, and N. Blakenbrough (Eds.), *Advances in Biochemical Engineering*, pp. 1-37. Springer-Verlag, Berlin, Heidelberg, New York, NY.

Torma, A. E., C. C. Walden, and D. W. Duncan. 1972. "The effect of carbon dioxide and particle surface area on the microbiological leaching of a zinc sulphide concentrate." *Biotechnol. Bioeng. 14*:777-786.

FEASIBILITY OF IN SITU CHEMICAL OXIDATION OF REFRACTILE CHLORINATED ORGANICS BY HYDROGEN PEROXIDE-GENERATED OXIDATIVE RADICALS IN SOIL

D. A. Martens and W. T. Frankenberger, Jr.

ABSTRACT

In situ application of hydrogen peroxide (H_2O_2) and ferrous (Fe^{2+}) salts (Fenton's reagent) for generation of hydroxyl radicals (OH·) appears to be a promising technology for decomposition of refractile organic contaminants. In a soil-free system, addition of Fenton's reagent (3 mL 2.79 M H_2O_2; 3 mL 0.1 M $FeSO_4$) to 2 mM *p*-chlorophenoxyacetic acid (372 μg *p*-CPA mL^{-1}) resulted in complete loss of the parent compound in 5 min as determined by high-performance liquid chromatography (HPLC) ultraviolet (UV) analysis with a stoichiometric release of Cl^- in 2 h. Treatment of 2,4-D (2 mM) and 2-[2,4,5-trichlorophenoxy]propionic acid (2 mM) with Fenton's reagent decomposed 72 and 86% of the parent compound in 5 h but failed to release stoichiometric levels of Cl^-. In soil systems, *p*-CPA was degraded from 2 to 15% in nontreated soils and from 31 to 86% after treatment with Fenton's reagent and 14 d of incubation at ambient temperatures as determined by free Cl^- analysis. Addition of Fenton's reagent decreased bacteria and fungal populations in the soils studied, but bacteria levels increased after 5 d to levels greater than the initial counts. The results suggest that Fenton's technology is compatible with bioremediation in contaminated soils.

INTRODUCTION

In nature, free radicals generated by photolysis or reaction of H_2O_2 and transition metals in the atmosphere or natural waters are nearly as ubiquitous as organic compounds themselves. The reaction of sunlight ($\lambda = 280$ to 430 nm) and ozone or nitrous acid in the atmosphere forming $OH\cdot$ is believed to be the mechanism responsible for degradation of nitric oxide and polycyclic aromatic hydrocarbons (PAHs) (Steinfeld 1989). Although $OH\cdot$ are highly reactive, their rates of production in natural waters (10^{-16} M) have been reported to be too low to react with organic contaminants (Mills et al. 1980). However, research has shown that generation of high levels of $OH\cdot$ by UV-ozone or H_2O_2 can successfully oxidize a wide range of organic contaminants in aqueous systems (Swallow 1978). The success of the UV oxidant (photolysis) is highly dependent on the clarity of the water. Cloudy or turbid waters have presented a problem with UV methods. No success has been reported for using UV-oxidant methods to decompose organic contaminants in soil.

However, application of $OH\cdot$ technology for in situ chemical oxidation may hold promise for destroying oxidizable contaminants in cloudy or turbid waters and also in soils. One mechanism for introducing strong oxidants into turbid waters or contaminated soils is the catalyzed decomposition of hydrogen peroxide to form $OH\cdot$:

$$H_2O_2 + Fe^{2+} \rightarrow OH\cdot + OH^- + Fe^{3+} \quad (1)$$

Fenton's reaction (Eq. 1) chemically generates $OH\cdot$, and the products generated include H_2O, Fe^{3+}, and O_2, which are environmentally benign and ubiquitous in natural waters and soils. The use of Fenton's reagent in aqueous systems has shown success for oxidation of chlorophenols (Barbeni et al. 1987), polychlorinated biphenyls (PCBs), chlorobenzene (Sedlak & Andren 1991a,b), chlorophenoxy herbicides (Pignatello 1992), formaldehyde (Murphy et al. 1989), sodium dodecylbenzene sulfonate (Sato et al. 1975), *p*-toluenesulfonic acid, *p*-nitrophenol (Feuerstein et al. 1981), and azo dyes (Kitao et al. 1982). Soil slurry systems treated with Fenton's reagent have also shown promise for decomposition of pentachlorophenol (Watts 1990) and trifluralin (Tyre et al. 1991).

Whereas in wastewater application the complete oxidation of the contaminant is often required, in soils it may be necessary to oxidize the hazardous component only to a biodegradable state. Therefore, the established goal with the use of Fenton's chemistry in aqueous systems would not be the same for soil matrices. This study was designed to determine the rates of H_2O_2 and Fe^{2+} addition required to decompose chlorinated compounds in soil and to determine if the use of Fenton's reagent is inhibitory to soil microorganisms.

MATERIALS AND METHODS

The four soils used were chosen to obtain a wide range in physiochemical characteristics (Table 1). The particle-size analysis was determined by a hydrometer method (Gee & Bauder 1986). Organic C was determined by a wet combustion method (Nelson & Sommers 1982). Total N was determined by a semi-Kjeldahl steam distillation method (Bremner & Mulvany 1982), and pH was determined on a 1:2.5 soil:water paste.

The chlorinated aromatic compounds were obtained from Sigma Chemical Co. (St. Louis, Missouri). Ferrous sulfate ($FeSO_4 \cdot 7H_2O$) and H_2O_2 were obtained from Fisher Scientific (San Francisco, California).

The decomposition of phenoxy herbicides in aqueous solution was determined by reacting 5 mL 1.5 M H_2O_2 and 5 mL 60 mM $FeSO_4$ with 4 mg *p*-chlorophenoxyacetic acid (*p*-CPA), 2,4-dichlorophenoxyacetic acid (2,4-D), or 2-[2,4,5-trichlorophenoxy]propionic acid (2,4,5-TP) at ambient temperatures for various times. The parent compound concentration was

TABLE 1. Properties of soils used.

Soil[a]	pH	Total C	Total N	Sand	Silt	Clay
		$g\ kg^{-1}$				
Arlington (Haplic Durixeralf)	7.9	2.30	0.20	670	80	250
Buren (Haplic Durixeralf)	7.3	5.20	1.00	575	150	275
Superstition (Typic Calciorthids)	6.0	0.38	0.17	925	50	25
Yolo (Typic Xerorthents)	8.4	5.00	1.19	575	200	225

[a] The Arlington and Buren soils were obtained from the Citrus Experimental Farm, University of California, Riverside, California; Superstition soil was obtained from the Yuma Experimental Station, Yuma, Arizona; and Yolo soil was obtained from the Vegetable Crops Experimental Farm, University of California, Davis, California.

determined by reverse-phase high-performance liquid chromatography (HPLC) using an R-Sil C_{18} 5-µm column (Alltech, San Jose, California) and isocratic elution with 60:40 methanol:water (0.5% acetic acid) with UV detection (254 nm). Peak identities were confirmed by cochromatography with authentic standards.

In this study, 10 g moist soil (oven-dry basis) were placed in a 125-mL Erlenmeyer flask and treated with 0 or 2 mM chlorinated aromatic compound. The reactive reagents were supplied by adding 3 mL containing 0 to 2.79 M H_2O_2 and 3 mL containing 0 to 0.1 M $FeSO_4$ in a 50-mL test tube. The mixture was added immediately to the contaminated soil, and the flask was swirled to mix the contents and incubated at ambient temperatures. After the specified time period, a water extract was obtained by adding 15 mL of H_2O to the flask and shaking the flask for 30 min at 4°C. The solution was filtered (Whatman #42) and the chloride released by the Fenton's treatment was determined with a coulometric titrator (Haake Buchler Instruments, Inc., Saddle Brook, New Jersey) and ion chromatography (Dionex Corp., Sunnyvale, California).

To monitor the effects of OH· generation on soil microbial populations, 10 g soil were treated with 3 mL 2.79 M H_2O_2 and 3 mL 0.1 M $FeSO_4$ for 1 h. A 1-mL aliquot of the treated soil-suspension was then taken for dilution plate counts using nutrient agar to enumerate the bacteria populations and Martin's rose bengal medium for fungal counts (Wollum 1982); counts were monitored daily.

RESULTS AND DISCUSSION

To determine if Fenton's reagent will degrade chlorinated phenoxyacetic herbicides, *p*-CPA, 2,4-D, and 2,4,5-TP were used as model compounds. Figure 1 shows an HPLC-UV chromatogram of *p*-CPA, 2,4-D, and 2,4,5-TP standards in aqueous solution. HPLC analysis determined that the parent *p*-CPA (372 µg mL^{-1}) was degraded by reaction with Fenton's reagent within 5 min and the system resulted in a stoichiometric release of Cl^- within 2 h as determined by ion chromatography. Additions of Fenton's reagent to 2,4-D and 2,4,5-TP (442 and 539 µg mL^{-1}, respectively) were less effective for degradation of these compounds when compared with decomposition of *p*-CPA. Figure 2 shows the concentration of 2,4-D and 2,4,5-TP before and after a 5-h reaction time with the Fenton's reagent. Approximately 72 and 86% of the 2,4-D and 2,4,5-TP parent compounds were decomposed by the OH· after 5 h of incubation, but the stoichiometric release of Cl^- was not achieved, suggesting that chlorinated aromatics and aliphatics remained in solution. The rapid and

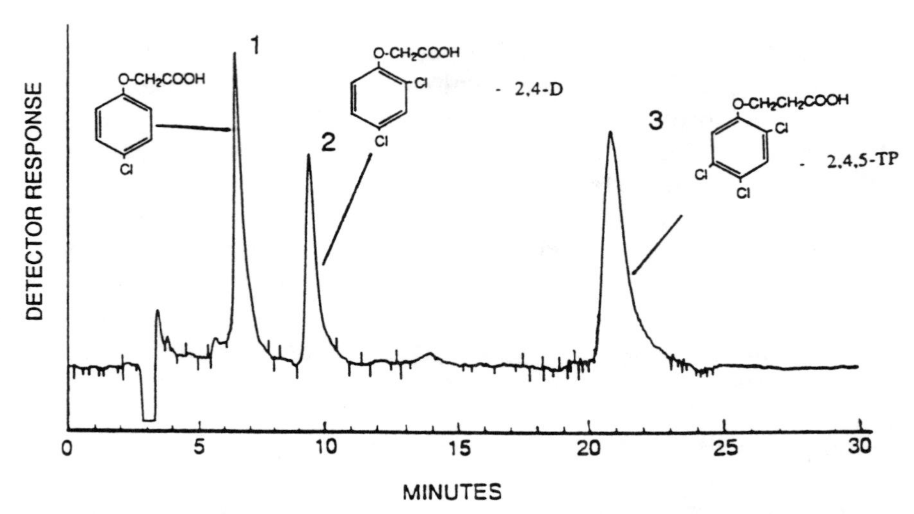

FIGURE 1. HPLC-UV chromatogram of phenoxy herbicide standards. 1 = *p*-chlorophenoxyacetic acid, 2 = 2,4-dichlorophenoxyacetic acid, 3 = 2-[2,4,5-trichlorophenoxy]propionic acid.

complete decomposition of *p*-CPA by the Fenton's reaction when compared with the decomposition of 2,4-D and 2,4,5-TP may be a factor of their water solubility. Of these, *p*-CPA (13.3 g L^{-1}) is more soluble than 2,4-D (0.90 g L^{-1}) or 2,4,5-TP (0.14 g L^{-1}). These results suggest that Fenton's reagent will degrade certain chlorinated aromatic compounds in aqueous solution and potentially may decontaminate organic wastes in soil by chemical oxidation, with possible decomposition of the resulting products by soil microflora.

In Fenton-treated soil it may be necessary to oxidize the hazardous waste only to a more biodegradable compound. Bioremediation in combination with chemical oxidation procedures would be a promising in situ technique if the OH· generated by the Fenton's reaction did not significantly reduce microbial populations in the treated soil. Table 1 shows the properties of the four soils used. The bacteria and fungal counts present in the soils with and without Fenton's addition are shown in Table 2. The results showed that addition of OH· via Fenton's reagent (3 mL 2.79 M H_2O_2; 3 mL 0.1 M $FeSO_4$) reduced both bacteria and fungal counts in the soils tested after treatment for 1 h. Continued sampling of the treated Yolo soil with time found that bacteria populations became greater than initial counts after 5 and 7 days incubation (Figure 3). The fungal counts did increase with time after treatment with Fenton's reagent but recovered to only one-half of the numbers determined initially (Figure 3).

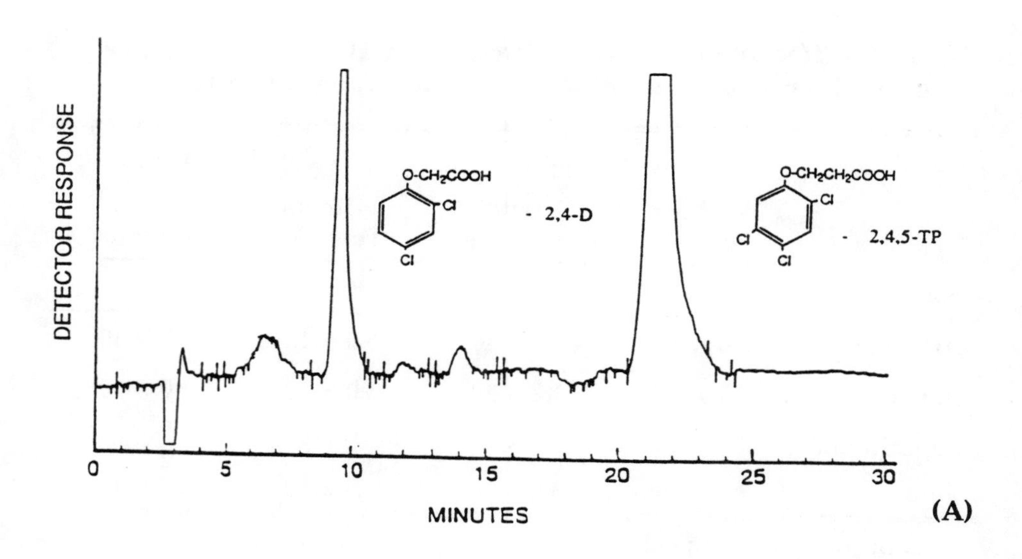

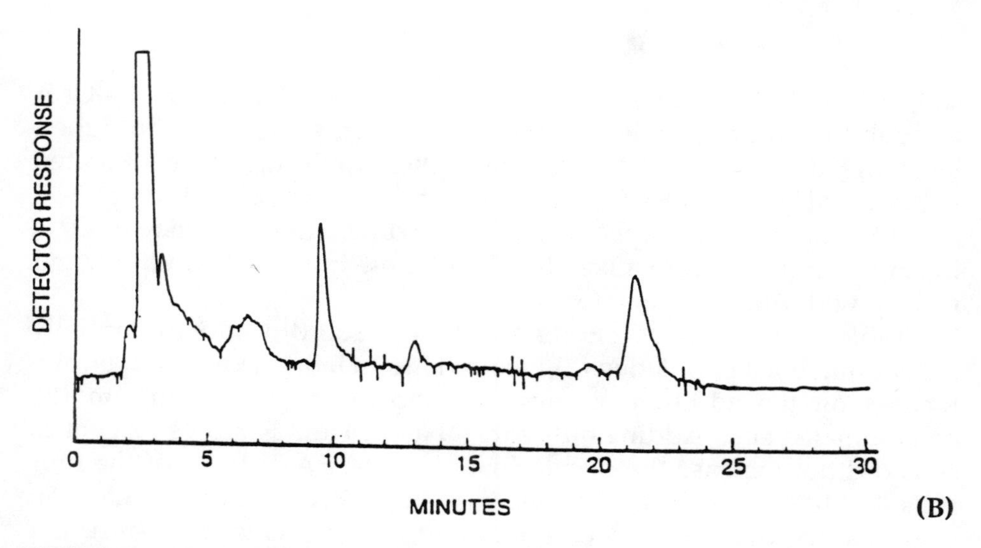

FIGURE 2. HPLC-UV chromatograms of (a) 2,4-D and 2,4,5-TP before reaction with Fenton's reagent, and (b) 2,4-D and 2,4,5-TP after reaction with Fenton's reagent for 5 h at ambient temperatures.

Previous work with acidified soil slurries (pH 3.0) has shown that addition of H_2O_2 without Fe^{2+} resulted in contaminant decomposition (Tyre et al. 1991, Watts 1990). Table 3 shows the effectiveness of different levels of H_2O_2 and Fe^{2+} for dechlorination of *p*-CPA in an Arlington soil (pH 7.9) that was not acidified to pH 3.0. The results show that in this

TABLE 2. Effect of Fenton's reaction (14.3 mg H_2O_2 and 1.39 mg $FeSO_4$ g^{-1} soil) on enumeration of bacteria and fungi in soils.

	Bacteria Counts		Fungi Counts	
	Without Fenton's	With Fenton's	Without Fenton's	With Fenton's
Soil	CFU/g soil[a]			
Arlington	5.0×10^7	5.4×10^4	6.0×10^5	2.5×10^3
Buren	3.1×10^6	1.2×10^5	6.0×10^5	1.0×10^5
Superstition	1.7×10^7	1.1×10^7	ND[b]	ND
Yolo	1.1×10^8	3.0×10^6	8.3×10^5	6.3×10^3

(a) CFU, colony-forming unit.
(b) ND, not detected.

soil, lack of Fe^{2+} additions was not effective for dechlorination of *p*-CPA and substantially higher dechlorination rates were achieved at the highest levels of H_2O_2 and Fe^{2+} additions. The highest application levels resulted in the initial H_2O_2 and $FeSO_4$ concentrations in the soil of 14.3 mg g^{-1} soil and 1.39 mg g^{-1} soil, respectively. To maximize the efficiency of OH·, the H_2O_2 and Fe^{2+} were mixed together before immediate addition and mixing with the contaminated soil.

Table 2 shows that Fenton's addition to soil did not substantially reduce microbial populations. Enhanced in situ bioremediation typically focuses on the addition of nutrients and an oxygen source to the contaminated site. Adding nutrients (N and P) and Fenton's reagent to soils did not increase the dechlorination of *p*-CPA in three of the four soils (Table 4). The lack of an increase in *p*-CPA by nutrient addition suggests that the natural levels of N and P in the soils were sufficient for mineralization of the contaminant. The addition of Fe^{2+} and urea peroxide (100 µg UP g^{-1} soil) was not as effective of a OH· source for the dechlorination of *p*-CPA, as was Fe^{2+} and H_2O_2 addition.

Many factors are involved in the potential success of degrading organic contaminants in soils. In the work presented here, Fenton's reagent rapidly decomposed a contaminant (*p*-CPA) with a higher water solubility than contaminants with much lower solubilities (2,4-D and 2,4,5-TP). The water solubility of the contaminant has been described as an important characteristic that is necessary for OH· attack (Sheldon & Kochi 1981). This study also found that high levels of Fenton's reagent

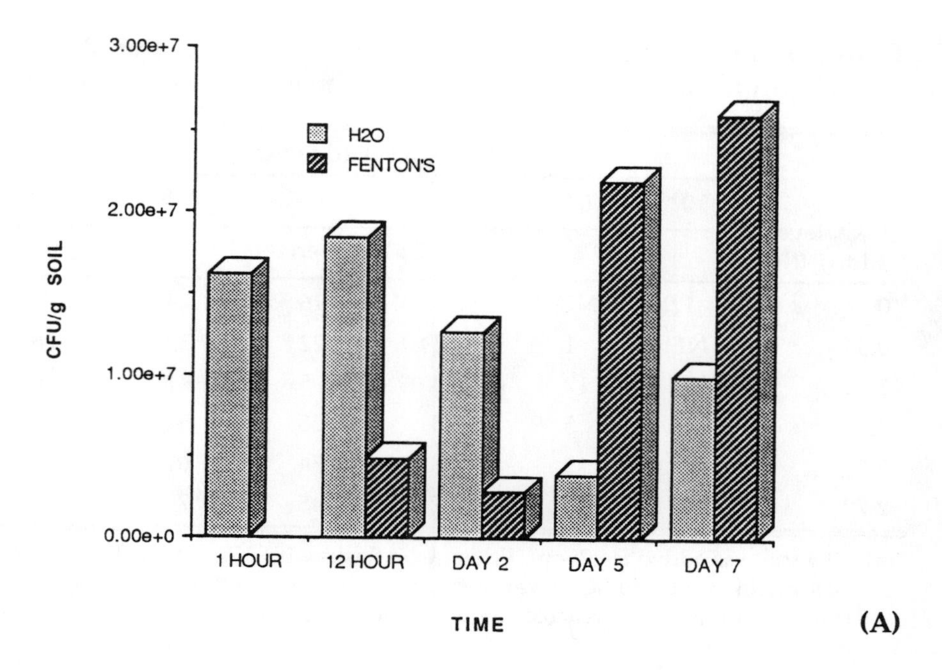

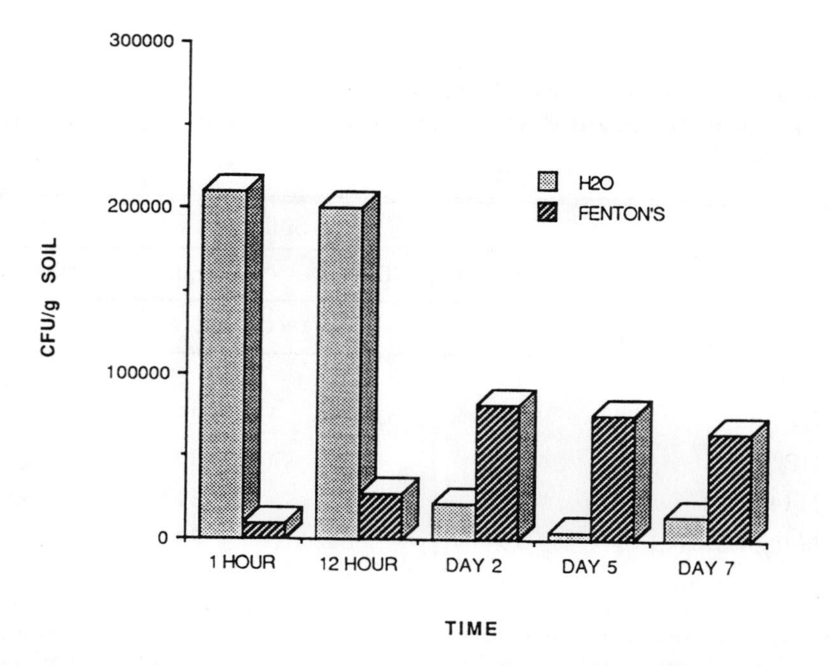

FIGURE 3. (A) Bacteria colony-forming units (CFUs) and (B) soil fungal CFUs with and without treatment of a Yolo soil with Fenton's reagent incubated at ambient temperatures for the specified times.

TABLE 3. Effects of different levels of H_2O_2 and $FeSO_4$ on dechlorination of *p*-CPA in an Arlington soil incubated for 2 weeks.[a]

	FeSO₄ level added (M)					
	0	0.02	0.04	0.06	0.08	0.10
H_2O_2 level added (M)	% Cl⁻ recovered					
0	1.9	ND[b]	ND	16	4	ND
0.56	ND	1	ND	22	39	26
1.12	9	17	4	5	41	50
1.67	15	5	23	28	52	60
2.23	6	15	28	34	61	67
2.79	2	19	37	45	75	83

(a) 10 g soil was added to 125-mL Erlenmeyer flasks with 3.72 mg *p*-chlorophenoxyacetic acid; 3 mL of H_2O at various concentrations and 3 mL $FeSO_4$ at various concentrations were reacted together immediately before addition to the contaminated soil.

(b) ND, not detected.

TABLE 4. Effects of Fenton's reagent or urea peroxide (UP) with nutrient addition on dechlorination of *p*-CPA in soils incubated for 2 weeks.[a]

	Soil			
	Arlington	Buren	Superstition	Yolo
Treatment	%Cl⁻ recovered			
Water only	15	9	2	9
Fe^{+2}; UP	47	18	23	13
Fe^{+2}; UP; P	53	31	20	11
Fe^{+2}; H_2O_2	86	85	31	50
Fe^{+2}; H_2O_2; Low N,P	79	84	23	35
Fe^{+2}; H_2O_2; High N,P	71	82	38	40

(a) 10 g soil contaminated with *p*-chlorophenoxyacetic acid (4 mg g^{-1} soil) was added to 125-mL Erlenmeyer flasks and treated with 3 mL Fe^{+2} (0.1 M) and 3 mL H_2O_2 (2.79 M) or urea peroxide (100 µg g^{-1} soil). Nitrogen (25 or 100 µg N g^{-1} soil) and phosphorus (10 or 40 µg P g^{-1} soil) were added, and the soil was incubated for 2 weeks.

addition (14.3 mg H_2O_2 and 1.39 mg $FeSO_4$ g^{-1} soil) did not eliminate soil microflora. The addition of N and P with Fenton's reagent did not substantially increase decomposition rates and the use of UP, as H_2O_2 was not as effective of a source of peroxide. Adsorption of the contaminant to soil organic matter is another important characteristic (Karickhoff 1981). OH· attack on soil organic material would also reduce the efficiency of Fenton's reaction by reducing the concentration of OH·. These factors must be addressed to determine the effectiveness of chemical oxidation as a means of decomposing persistent organics in soil.

ACKNOWLEDGMENTS

This project has been funded by the Kearney Foundation of Soil Science. The authors would like to thank Ms. Stephanie Luck for her valued laboratory assistance.

REFERENCES

Barbeni, M., C. Minero, and E. Pelizzetti. 1987. "Chemical Degradation of Chlorophenols with Fenton's Reagent (Fe^{2+} + H_2O_2)." *Chemosphere 16*:22-25.

Bremner, J. M., and C. S. Mulvaney. 1982. "Nitrogen-Total." In A. Page, R. H. Miller, and D. R. Keeney (Eds.), *Methods of Soil Analysis. Part 2. Chemical and Microbiological Properties*, pp. 595-622. American Society of Agronomy, Madison, WI.

Feuerstein, W., E. Gilbert, and S. H. Eberle. 1981. "Model Experiments for Oxidation of Aromatic Compounds by Hydrogen Peroxide in Wastewater Treatment." *Vom Wasser 56*:35.

Gee, B. W., and J. W. Bauder. 1986. "Particle-Size Analysis." In A. Klute et al. (Eds.), *Methods of Soil Analysis. Part 1. Physical and Mineralogical Methods*, pp. 399-404. American Society of Agronomy, Madison, WI.

Karickhoff, S. W. 1981. "Semi-Empirical Estimation of Sorption of Hydrophobic Pollutants on Natural Sediments and Soils." *Chemosphere 10*:833-846.

Kitao, T., Y. Kiso, and R. Yahashi. 1982. "Studies on the Mechanism of Decolorization with Fenton's Reagent." *Mizii Shori Gijutsu 23*:1019.

Mills, T., D. G. Hendry, and H. Richardson. 1980. "Free-Radical Oxidants in Natural Waters." *Science 207*:886.

Murphy, A. P., W. J. Boegli, M. K. Price, and C. D. Moody. 1989. "A Fenton-Like Reaction to Neutralize Formaldehyde Waste Solutions." *Environmental Science and Technology* 23:166.

Nelson, D. W., and L. E. Sommers. 1982. "Total Carbon, Organic Carbon, and Organic Matter." In A. Page, R. H. Miller, and D. R. Keeney (Eds.), *Methods of Soil Analysis. Part 2. Chemical and Microbiological Properties*, pp. 539-577. American Society of Agronomy, Madison, WI.

Pignatello, J. J. 1992. "Dark and Photoassisted Fe^{3+}-Catalyzed Degradation of Chlorophenoxy Herbicides by Hydrogen Peroxide." *Environmental Science and Technology* 26:944.

Sato, S., T. Kobayashi, and Y. Sumi. 1975. "Removal of Sodium Dodecylbenzene Sulfonate with Fenton's Reagent." *Yukagaku* 24:863.

Sedlack, D. L., and A. W. Andren. 1991a. "Oxidation of Chlorobenzene with Fenton's Reagent." *Environmental Science and Technology* 25:777.

Sedlack, D. L., and A. W. Andren. 1991b. "Aqueous-Phase Oxidation of Polychlorinated Biphenyls by Hydroxyl Radicals." *Environmental Science and Technology* 25:1419.

Sheldon, R. A., and J. K. Kochi. 1981. *Metal-Catalyzed Reactions of Organic Compounds: Mechanistic Principles and Synthetic Methodology Including Biochemical Processes*. Academic Press, New York, NY.

Steinfeld, J. H. 1989. "Urban Air Pollution: State of the Science." *Science* 243:745.

Swallow, A. J. 1978. "Reactions of Free Radicals Produced from Organic Compounds in Aqueous Solution by Means of Radiation." *Progressive Reaction Kinetics* 9:195.

Tyre, B. W., R. J. Watts, and G. C. Miller. 1991. "Treatment of Four Biofractory Contaminants in Soils Using Catalyzed Hydrogen Peroxide." *Journal of Environmental Quality* 20:832.

Watts, R. J. 1990. "Treatment of Pentachlorophenol-Contaminated Soils Using Fenton's Reagent." *Hazardous Waste and Hazardous Materials* 7:335.

Wollum, A. G. 1982. "Cultural Methods for Soil Microorganisms." In A. Page, R. H. Miller, and D. R. Keeney (Eds.), *Methods of Soil Analysis, Part 2. Chemical and Microbiological Properties*. pp. 781-802. American Society of Agronomy, Madison, WI.

IMPLEMENTATION OF MICROBIAL MATS FOR BIOREMEDIATION

J. Bender and P. Phillips

ABSTRACT

Constructed microbial mats, composed of heterotrophic and autotrophic microbes, have been developed for various bioremediation applications. These mats, cultured on ensiled grass clippings, are durable, highly resilient, self-sustaining ecosystems with few growth requirements. Mats demonstrate the following characteristics, which are useful in bioremediation: (1) contain oxidizing and reducing zones; (2) raise pH of their aqueous environment; (3) generate high oxygen and redox levels during the photoperiod; (4) release negatively charged bioflocculents; and (5) self-attach to a variety of substrates, thereby immobilizing the total ecosystem. Mats have been found to reduce selenate to elemental selenium; remove Pb, Cd, Cu, Zn, and Mn from water; and degrade TNT, chlordane, chrysene, naphthalene, and phenanthrene. Mat treatment of petroleum distillates shows evidence of complete mineralization. The diverse microbial components within the mat define the range of molecular, cellular, and communal mechanisms available to this ecosystem and likely account for the broad range of successful bioremediation applications recently demonstrated with this system.

INTRODUCTION

Mixed microbial mats, composed of stratified layers of microbes, are resilient, primeval communities dominated by cyanobacteria (blue-green algae). Historically, they have occupied the most inhospitable environments on earth (Ward et al. 1989). Cyanobacteria, which fix both carbon and nitrogen, provide the base of support for heterotrophic bacteria that colonize the lower regions (Paerl et al. 1989). Purple autotrophic bacteria typically occupy the region below the cyanobacteria where light intensities and oxygen concentrations are low. This self-organizing consortium of microbes forms a laminated, multilayered biofilm in water, sediments, and soils. These microbes can be generated on water surfaces by enriching shallow water with ensiled grass clippings (Bender et al. 1989, 1991a). Once established, the mat becomes annealed together by a gel matrix and persists for long periods without nutrient supplements.

Intact mats, because of their functional diversity, may bring a flexibility and efficiency to biotechnology that is frequently lacking in single-species systems. Because mixed microbial mats evolve under hostile conditions, similar to those expected in highly contaminated environments, survival adaptations of these ecosystems are directly applicable to remediation biotechnology. Additionally, the properties of self-maintenance and resilience under fluctuating environmental conditions may resolve a number of maintenance problems often associated with bioremediation technologies.

This paper reviews earlier published work from our laboratory on the characteristics of mats and their applications to the sequester of heavy metals (Pb, Cd, Cu, Zn) and the reduction of selenate to elemental selenium. In addition, new data are presented relating to on-site remediation of acid mine drainage and laboratory-scale research in the biodegradation of recalcitrant organics such as petroleum distillates, TNT, and chlordane. Because environmental pollution problems are often compounded by the occurrence of heterogeneous mixtures of heavy metals, metalloids, and organic contaminants, the capacity of a system to remediate various categories of contaminants is clearly advantageous.

Procedures for new research are described, but detailed methods from previously published work are omitted. All applications of mats in various remediation problems from previous work and current research are presented and compared.

MATERIALS AND METHODS

Development of Mats for Bioremediation

The mat consortium, containing bacteria and cyanobacteria, is developed by constructing an artificial ecosystem of a soil base, a water column, and a floating layer of ensiled grass clippings in laboratory trays (Bender et al. 1989). Purple sulfur-oxidizing bacteria (*Chromatium* spp.) were introduced to remove the H_2S in anoxygenic photosynthesis and thereby lower the reducing power of the system. As necessary, mats were adapted to the target contaminants by stepwise exposure to increasing concentrations.

Finished mats were dried for use as inocula in the field and laboratory experiments.

Methods for Mine Drainage Treatment

Pond Construction. Figure 1 illustrates the design for the treatment of acid coal mine drainage contaminated with Fe and Mn. Three ponds, prepared for biological treatment were lined with PVC film. Limestone rocks (2 to 3 cm) were added in two of the ponds and pea gravel (1 to 2 cm) in the third. Rocks were layered in such as way as to construct alternating high and low regions ranging from 2 to 30 cm in depth, spaced 1 m apart.

Development of Inocula for Ponds. Microbial strains (including *Oscillatoria* spp., green filamentous algae, and *Chromatium* spp.) were

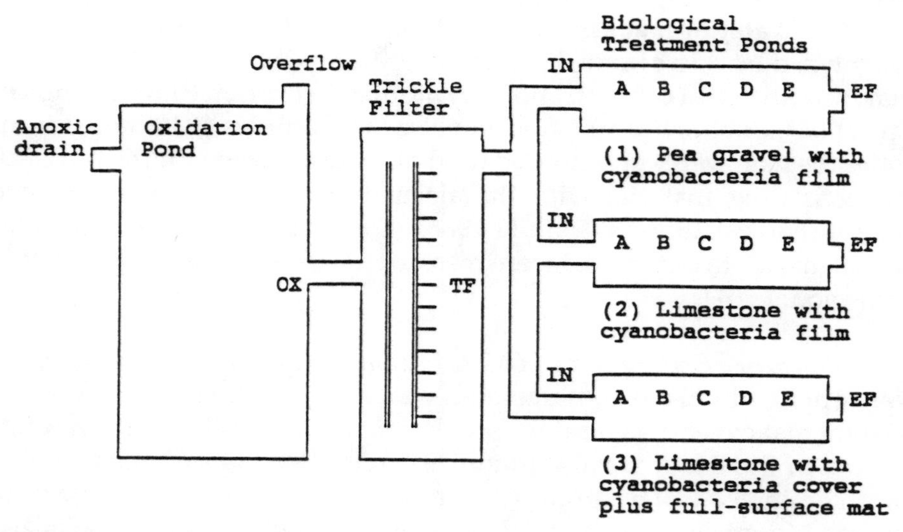

FIGURE 1. Schematic of the integrated treatment system designed for iron and manganese removal from acid mine drainage. *Legend.* Oxidation pond = approximately 1 ha, trickling filter = 20 m^2; biological treatment ponds (1) = 32 m^2, (2) = 44 m^2, (3) = 44 m^2. Sampling points are OX, oxidation pond near effluent; TF, center of trickling filter (water is dripped through two delivery pipes to lower iron concentration); IN, influent water from trickling filter; EF, effluent water from biological treatment ponds; A-E sample points have the following distances from inflow (m): 0.3, 1, 2, 5, and 8.

selected from the treatment area and developed into silage-microbial mats in the laboratory. Mature laboratory mats (composed of microbes initially selected from the target site) were used as field inocula for the ponds. Mixtures of mats and silage were prepared and broadcast over the pond at rate of 1 to 1.5 L/pond in three applications during a 4-week period. After the mat covered approximately 30% of the experimental pond surface, the drainage water was delivered to the three ponds. Flowrates were increased approximately every 6 weeks from 1 to 2.5 to 5 L/min. Because of the difficulty of maintaining total absence of biomass in field ponds, controlled experiments for metal uptake by clean rocks were performed in the laboratory.

Analysis of Ecosystem Parameters in Field Ponds. Pond sampling points (2 cm depth) were measured for DO (Otterbine Sentry III dissolved oxygen meter), pH (Orion series 200 with GX series electrode), Eh (Orion 200 series meter with platinum redox combination electrode), conductivity (Fisher Scientific digital conductivity meter), and concentrations of Fe and Mn. Water or mat samples for all metal analysis experiments were hydrolyzed by microwave digestion (CEM Corp. Model MDS-2000) and analyzed for metal concentrations by atomic absorption (Varian, Spectra AA-20 BQ, double beam). Additional water samples (1 L), taken from under the mat, were concentrated to 35 mL and analyzed for the presence of flocculating material with the alcian blue assay for algal anionic polysaccharides (Bar-Or & Shilo 1988). The surface charge of the bio-flocculents (Zeta potentials) were measured with a Laser Zee meter (Pen Kem model 501).

Laboratory Experiments in Metal Sequestering. Metal-tolerant mats were applied to metal-contaminated water in three treatment methods: (1) total mat cover of laboratory ponds containing a soil bed and a water column of 1 to 3 L of metal solution, (2) excised sections of mats applied to contaminated water samples (9 cm^2 mat per 0.2 L of field sample from the Iron Mountain Mine drainage in California), and (3) mats immobilized on 4-g glass wool floating balls (floaters) and applied to a mixed solution of Zn/Mn-contaminated water.

Methods for Biodegradation Experiments

TNT Degradation. Methods of mat development for TNT degradation and analysis are those of Mondecar et al. (1992). Metal-tolerant mats were integrated with TNT-tolerant bacteria isolated from a contaminated site. Finished mats were incubated with up to 100 mg/L TNT, solubilized by overnight shaking. Water columns and mats were analyzed for residual TNT and metabolites. Degradation was examined by an NaOH plate assay (Osmon & Klausmeier 1972) and by high-pressure liquid chromatography (HPLC) (Perkin-Elmer C-18 reverse-phase column with methanol/water (1:1) as the mobile phase (flowrate of 1 mL/min). Metabolites were derived with trifluoroacetic acid anhydride. Compounds were detected with a LC85B Perkin-Elmer Ultraviolet Detector at 254 nm.

Petroleum Degradation. In a light and dark series, microbial mineralization (measured as CO_2 levels absorbed by a KOH trapping agent) of ^{14}C-labeled hexadecane and chrysene and phenanthrene was monitored by liquid scintillation counting to 90 days. In a light and a dark series of flasks, mat "plugs" (1.00 to 1.69 g) were placed in 250-mL acid-washed glass flasks containing 100 mL of Allen and Arnon Modified Media (1955). These were sealed with a Teflon™-coated stopper. The ^{14}C-phenanthrene, chrysene (Amersham/Searle Corp., Arlington Heights, IL), and hexadecane (Sigma Chemical Co., St. Louis, MO) were spiked at greater than 4,800 dpm/mL/flask in a triplicate experimental series.

Chlordane Degradation. Mats were developed for chlordane tolerance by incubating excised sections with increasing concentrations of chlordane. Chlordane was added to the water column under floating mats at a concentration of 2,100 mg/L. Although chlordane is not soluble in water, the mats produced biofilms that actively sequestered the globules of chlordane from the water column and transported them into the mat. This could be visually observed. In a second experiment, chlordane was mixed in soil at a concentration of 200 mg/kg. Contaminated soil was packed in test tubes (wrapped to maintain darkness in the soil and light at the mat surface). Mats were cultured on the moist chlordane-soil (without water column) for 25 days. All experimental mats, water columns, and soils were removed and extracted according to U.S. Environmental Protection Agency (1986) procedures and analyzed for chlordane by Woodson-Tenent Laboratories, Gainesville, Georgia (using a Tracor 540 gas chromatograph with electron capture detector).

RESULTS AND DISCUSSION

Mine Drainage Treatment

A floating mat (1 to 2 cm thick), composed of filamentous green algae and cyanobacteria, predominantly *Oscillatoria* spp., grew rapidly in the pond after addition of silage-microbial mat inocula. A secondary mat of cyanobacteria also covered the limestone at the bottom. Thus the metal-contaminated water flowed between the double-layered mats. Approximately 6 weeks were required to establish a full pond mat cover, but effective metal removal began in the early stages of mat growth.

The two ponds designed as controls rapidly became inoculated with an *Oscillatoria* strain resembling that of the inoculated cyanobacteria. This strain formed a thin layer (<0.5 mm) on the rocks. No floating mat developed in these two ponds, and the biomass remained relatively low compared to that in the experimental pond. Table 1 and Figure 2 show the metal removal profiles and water conditions in the ponds at a flowrate of 2 to 5 L/min during light and dark periods. Metals were effectively removed after the inflow water had flowed a distance of approximately 1 to 2 m through the pond. All sampling points beyond that distance (points C to the outflow) showed metal concentrations <1.2 mg/L.

The two ponds with only *Oscillatoria* spp. films on the rocks also showed metal removal. However, metals were removed more slowly in these ponds, most noticeably at night (Figure 2-B). Laboratory controls with limestone only showed a 25% Mn removal. These laboratory experiments were added because it was impossible to keep microbial films from contaminating the field pond limestone.

The floating mat in the experimental pond remained healthy and showed no signs of metal deposit on its surface except for iron hydroxide precipitates near the influent. Black deposits of Mn were generally deposited between the two mat layers and did not impact the biological activity of the mat. Redox and DO levels were high during the light period, and pH levels ranged from 6.4 to 7.7. Even after 10 h of darkness, oxygen levels remained at 6 mg/L in some regions of the mat pond. During the photosynthetic period, bubbles of oxygen become entrapped in the slimy matrix that typically binds the mat. Apparently this sequestered oxygen remains available throughout the night. This is to be expected, because cyanobacteria are unusual in that they have limited ability to utilize organic substrates for energy production in the dark (Kratz & Myers 1955); thus the oxygen consumption remains low in this pond.

Although the conditions of high oxygen and high Eh may be central to the deposit of Mn oxides, other factors may be functional as well. Flocculents were identified in the water column under the mat. Laboratory research showed that specific bioflocculents were released by the mat in response to the presence of Mn^{+2} (Rodriguez-Eaton et al. 1994). These materials carried surface charges ranging from −58.8 to −65.7 mV. The charges changed to +1.8 in the presence of divalent metal, indicating metal binding to the bioflocculent.

TABLE 1. Environmental parameters and metal concentrations measured in the three-phase field pond treatment system designed for Fe and Mn removal.

Sample	DO	pH	Eh, mV	Conduct.	Metal conc., mg/L	
					Mn	Fe
Day Samples						
OX	5.7	6.6	430	703	9.7	11.9
TF	7.0	nm	nm	nm	5.1	2.4
Pond: CGM						
IN	6.1	6.6	469	689	4.5	16.3
A	4.9	6.7	401	708	4.3	3.2
B	8.0	6.8	444	702	nm	1.9
C	7.2	6.9	474	725	0.5	1.1
D	10.5	7.2	426	778	0.1	0.6
E	16.0	7.7	438	775	0.1	0.4
Night Samples						
OX	7.2	7.0	409	694	9.8	4.7
TF	8.0	7.0	407	698	8.3	10.2
Pond: CGM						
IN	6.0	6.9	428	649	3.6	4.4
A	6.0	6.8	463	680	3.4	1.0
B	4.5	6.9	461	685	1.2	0.5
C	4.0	7.2	460	681	0.4	0.2
D	3.0	7.2	454	740	0.2	0.2
E	2.0	7.2	447	715	<0.1	<0.1

Sample points are identified in Figure 1. OX, oxidation pond; TF, trickling filter; IN, influent water; A-E, sample points in biological treatment pond; CGM, experimental pond containing a floating mat of cyanobacteria/green algae mat (1 to 2 cm thick) and a bottom mat of cyanobacteria (1 to 2 mm thick) covering limestone rocks. Influent water was delivered between the two mats at a flowrate of 2 to 5 L/min. TF day samples were taken on a different day from remaining samples. All data are from unfiltered samples. Oxidation pond and trickling filter samples contained iron precipitates that were removed by filtering. nm = not measured.

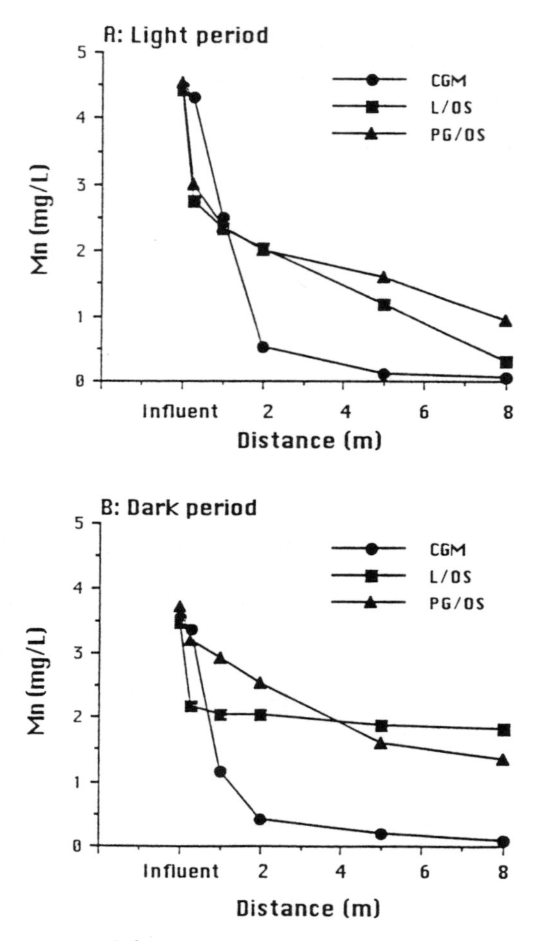

FIGURE 2. Mn removal from acid mine drainage. *Legend.* Sampling points are illustrated in schematic of Figure 1. CGM is the limestone substrate pond with a cyanobacteria and green algae mat, L/OS is the limestone substrate pond and PG/OS is the pea gravel substrate pond. Both became colonized with cyanobacteria. Graph A = light period (11:00 am); B = dark period (6:00 am). Control experiments (limestone only, performed in the laboratory) showed approximately 25% Mn removal at a 3-m flow distance.

No soil was layered in the pond, so the predictable microbial ecology characterizing the sediment region may not be present in this system. The primary mechanisms of deposit likely were determined primarily by the chemical/biological processes mediated by the mat. Metals are known to complex with a wide range of organic materials, including microbes and their organic releases. Dunbabin and Bowmer (1992) identify four dominant binding processes that incorporate metals into organic materials: (1) cation exchange, (2) adsorption, (3) precipitation/co-precipitation, and (4) complexation or chelation.

Although metals that are adsorbed, precipitated, or complexed can be released back into solution in an equilibrium response, no such fluctuations have been detected thus far during a 4-month experimental period. Additionally, conditions of neutral pH with high DO and redox levels (mediated by the biology of the mat ecosystem) favor the chemical precipitation of Mn oxides and Fe hydroxides. These, in turn, act as reservoirs for additional metal deposits. Also, the pH, redox, and oxygen levels mediated by the mat provide the environmental conditions for bacterial oxidation of Mn by heterotrophic populations colonizing with the mat.

The potential bioavailability of metals is favored by increases in acidity, reducing power, and salinity (Dunbabin & Bowmer 1992). Constructed mats, containing *Oscillatoria* spp. and *Chromatium* spp., would tend to lower bioavailability by raising the pH and redox of the system.

Although anaerobic zones have been identified within the laboratory-cultured mats (Bender et al. 1989) and are likely present in the field mat, the redox conditions of the water column under the mat in the experimental pond remained high even after extended dark periods (Table 1).

Laboratory Experiments in Metal and Metalloid Removal. Table 2 gives a summary of metal removal with several types of mat applications in quiescent laboratory ponds. Mats immobilized on glass wool substrate showed good potential for metal removal. Although the metal removal rate of excised mats placed in concentrated metal solution (Iron Mountain Mine sample) was comparatively high, the mats stopped growing and showed signs of cell death. Therefore, this method of treatment may be best for single-batch exposures, rather than for continuous treatment.

It is likely that several mechanisms, including flocculation, cell sorption, and mediation of the environmental conditions of pH, Eh, and oxygen, contribute to metal removal. In immobilized and free-floating mat treatments, most metal deposits as well as elemental selenium (product of selenate reduction by the mat) remained associated with the floating mat and were removed simply by raking the biomass from the surface of the ponds (Bender et al. 1989, 1991a). Electron micrographs, made from the metal-exposed cyanobacteria, suggest that most metal is deposited outside of the cells. This fact may explain the durability of the mats during metal removal.

Metal-packing levels in the biomass varied according to exposure concentration, duration of exposure, and quantity of mat. The highest levels were achieved by mats immobilized on glass wool floaters. In this system organic material, released from the mats and entrapped in the glass wool floater, showed packing levels of 380 and 240 mg/g for Zn and Mn, respectively. The mat itself bound 35 and 40 mg/g Zn and Mn. Clearly, the mat releases, which likely include the negatively charged bioflocculents, are important in binding high concentrations of metals. Mats showed no metal toxicity at these binding levels.

Biodegradation by Mats

Degradation of TNT. Mats were constructed specifically for TNT degradation by incubating TNT-tolerant bacteria (isolated from the Bangor

TABLE 2. Metal and metalloid removal in quiescent laboratory ponds.

Treatment system	Initial concentration, mg/L	Removal rate, mg metal/m^2 mat/h
Free floating mats	Pb: 117	129
	Se: 37	6
Mat immobilized on floaters	Mix of	
	Zn: 22	313
	Mn: 18	462
Excised mats	Mix of	
	Cu: 284	378
	Zn: 3,021	3,778
	Cd: 19	356

Free floating mat. Self-buoyant mats were cultured on the surface of laboratory ponds containing Pb or Se. Initial solution of selenate was reduced in part to elemental selenium which deposited in the surface mat (Bender et al. 1991a). Pb was deposited in the mat as PbS (Bender et al. 1989). The pH for the free floating mats was 6 to 8.

Mat immobilized on floaters. The mat was attached to glass wool balls that were floated in Zn/Mn-contaminated water at pH 7 to 9 (Bender 1992b).

Excised mats. Small sections of mat were excised and applied to a mixed solution of Cu, Zn, Cd, and Fe sample from Iron Mountain Mine drainage in California (Bender et al. 1991b). The pH was adjusted to 3 to 4 before adding mat sections. nm = not measured.

Naval Submarine Base, Washington) with the mat consortium. Earlier experiments indicated that simple incubation resulted in an integration of the desired bacteria into the mat. Although the bacteria applied as a single species to the TNT showed a degradation rate of 0.78 mg/d, integrated mats (prepared by pre-incubation with TNT-tolerant bacteria) degraded TNT at a rate of 3.3 mg/d. Expected degradative products of 2-amino-4, 6-dinitrotoluene and 4-amino-2, 6-dinitrotoluene appeared only in low concentrations (<6 mg/L). These metabolites generally decreased as mat exposure time increased. No other metabolites were found.

Mineralization of Petroleum Distillates. The $^{14}CO_2$ measurements in the light-series experiments likely were confounded by possible $^{14}CO_2$ re-incorporation by the photosynthetic cyanobacteria. After 90 days, dark-series total mineralization levels were 24.1% for phenanthrene, 20.5% for chrysene, and 9.3% for hexadecane (Table 3 and Figure 3).

The specific microbial components active in hexadecane and chrysene degradation were further investigated. Using six constructed silage-microbial mats (of known microbial character), during 28 days under light conditions, the mat wet weight increased an average of 103.3% in

TABLE 3. Biodegradation by constructed microbial mats.

Contaminant	Concentrations, mg/L		Time and percent degradation
	Initial	Final	
2,4,6-trinitrotoluene (TNT)	100	<1	>99% in 6 days
Chlordane			
in water	2,100	61	97% in 10 days
in soil	200	146	27% in 25 days
Petroleum distillates[a]			
hexadecane	768	697	9% in 90 days
phenanthrene	374	284	24% in 90 days
chrysene	157	125	20% in 90 days

(a) Percent degradation for petroleum distillates designates mineralization in dark-cycle experiments (determined by ^{14}C-labeled carbon dioxide collected in KOH traps). Quantities of mat applied to the substrates (mat surface area, cm^2) were TNT = 2 per petri plate, chlordane in water = 16.0 per 50 mL media, chlordane in soil = 2.5 per test tube, and petroleum distillates = 2.0 per 100 mL media.

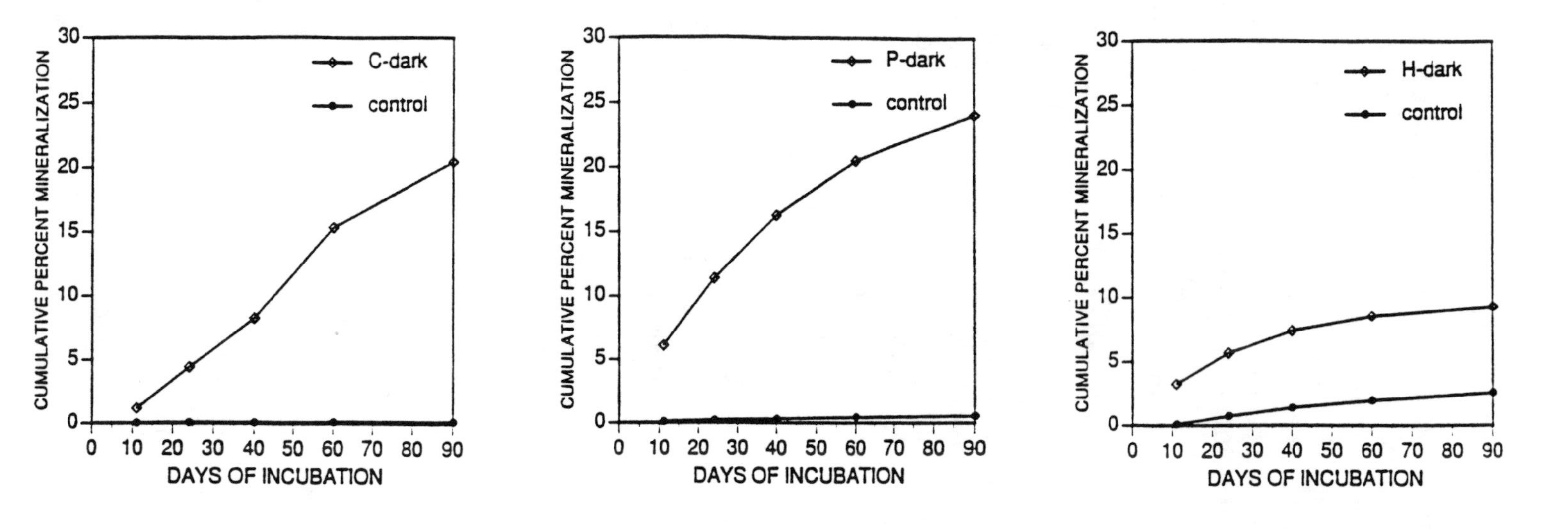

FIGURE 3. Mineralization of petroleum distillates by microbial mats during dark period. *Legend.* C: chrysene, P: phenanthrene, H: hexadecane.

hexadecane and 141% in chrysene. Mat wet weights generally decreased in the dark-series experiments in hexadecane (mean = –7.2%) and chrysene (mean = –26.0%). After 28 days, the light-series daily mineralization rates ranged from 0.022 to 0.644 ng/d for chrysene to 3.5 to 100.4 ng/d for hexadecane. Dark-series values for the same time period ranged from 0.494 to 2.998 ng/d for chrysene and 63.2 to 1232.8 ng/d for hexadecane. These data indicate that consortial activities of several microbial members are operative in the degradation of the distillates.

Degradation of Chlordane. The mats adapted for chlordane tolerance showed 97% degradation of 2,100 mg/L chlordane in 10 days (Table 3). Chlordane mixed in soil (200 mg/kg) was 27% degraded in 25 days. Motile *Chromatium* spp., which generally colonizes under the mat, rapidly migrated through the chlordane-contaminated soil. Because *Chromatium* spp. has been identified as one of the microbial mat members that degrades chlordane (Bender 1992a), it is likely that this group was active in soil decontamination.

CONCLUSIONS

This microbial mat system demonstrates good potential in metal sequestering and degradation of recalcitrant organics. Although the molecular, cellular, and communal mechanisms available in the mats are elegantly complex, the implementation of the system is simple and the results are highly reproducible. Our future mat research will focus on simultaneous remediation of mixed contaminants.

ACKNOWLEDGMENTS

Infrastructure support for all laboratory research was provided by U.S. Environmental Protection Agency Grant No. CR81868901 to Clark Atlanta University. On-site treatment of acid mine drainage was supported by the Tennessee Valley Authority Contract No. TV-89721V.

REFERENCES

Allen, M. B., and D. I. Arnon. 1955. "Studies on Nitrogen-Fixing Blue-Green Algae 1. Growth and Nitrogen Fixation by *Anabaena cylindrica*." *Lemm. Pl. Physiol. 30*: 366-372.

Bar-Or, Y., and M. Shilo. 1988. "Cyanobacterial Flocculants." *Meth. Enzymol. 167*: 616-622.

Bender, J. 1992a. "Bioremediation Potential of Microbial Mats." Abstract presented at Bioremediation Workshop. Office of Naval Research, Seattle, WA.

Bender, J. 1992b. "Recovery of Heavy Metals with a Mixed Microbial Ecosystem." Final report to the U.S. Bureau of Mines, Grant No. G0190028.

Bender, J. A. 1991. "Reclamation of Metals with a Silage-Microbe Ecosystem." In *Pollution Abatement and Installation Restoration Research and Development*. Research Summary Report Prepared by Science and Technology Corporation for U.S. Army Toxic and Hazardous Materials Agency, Aberdeen Proving Ground, MD.

Bender, J., E. R. Archibold, V. Ibeanusi, and J. P. Gould. 1989. "Lead Removal from Contaminated Water by a Mixed Microbial Ecosystem." *Water and Science Technology 21/12*: 1661.

Bender, J., J. P. Gould, Y. Vatcharapijarn, and G. Saha. 1991a. "Uptake, Transformation and Fixation of Se(VI) by a Mixed Selenium-Tolerant Ecosystem." *Water, Air, and Soil Pollution 59*:359.

Bender, J., J.P. Gould, and Y. Vatcharapijarn. 1991b. "Sequester of Zinc, Copper and Cadmium from Water with a Mixed Microbial Ecosystem." *Proceedings of the U.S. Army USATHEMA Conference*, Williamsburg, VA.

Dunbabin, J. S., and K. H. Bowmer. 1992. "Potential Use of Constructed Wetlands for Treatment of Industrial Wastewaters Containing Metals." *The Science of the Total Environment 111*: 151.

Kratz, W. A., and J. Myers. 1955. "Photosynthesis and Respiration of Three Blue-Green Algae." *Pl. Physiol., Lancaster 30*:275.

Mondecar, M., J. Bender, J. Ross, W. George, and K. Dummons. 1992. "Bioremediation of TNT by a Mixed Microbial Ecosystem." Abstract presented at *92nd Annual Meeting of the American Society for Microbiology*, Washington, DC.

Osmon, J. L., and R. E. Klausmeier. 1972. "The Microbial Degradation of Explosives." *Dev. Ind. Microbiol. 14*: 247-252.

Paerl, H. W., B. M. Bebout, and L. E. Prufert. 1989. "Naturally Occurring Patterns of Oxygenic Photosynthesis and N_2 Fixation in a Marine Microbial Mat: Physiological and Ecological Ramifications." In Y. Cohen and E. Rosenberg (Eds.), *Microbial Mats*, pp. 326-341. American Society for Microbiology, Washington, DC.

Rodriguez-Eaton, S., U. Ekanemesang, and J. Bender. 1994. "Release of metal-binding flocculents by microbial mats." In J. L. Means and R. E. Hinchee (Eds.), *Emerging Technology for Bioremediation of Metals*. Lewis Publishers, Ann Arbor, MI.

U.S. Environmental Protection Agency. 1986. Test Method for Evaluating Solid Waste Laboratory Manual 846. 3540-7. EPA. Washington, DC.

Ward, D. M., R Weller, J. Shiea, R. W. Castenbolz, and Y. Cohen. 1989. "Hot Spring Microbial Mats: Anoxygenic and Oxygenic Mats of Possible Evolutionary Significance." In Y. Cohen and E. Rosenberg (Eds.), *Microbial Mats*, pp. 3-15. American Society for Microbiology, Washington, DC.

MERCURY MICROBIAL TRANSFORMATIONS AND THEIR POTENTIAL FOR THE REMEDIATION OF A MERCURY-CONTAMINATED SITE

E. Saouter, R. Turner, and T. Barkay

INTRODUCTION

Methylmercury ($CH_3Hg(I)$) is the chemical form of mercury that is bioconcentrated by aquatic biota, and thus may constitute a threat to public health (D'Itri & D'Itri 1978). Remedial intervention based on degradation of $CH_3Hg(I)$ and removal of the substrate for mercury methylation (inorganic mercury; Hg(II)) would result in a decreased concentration of $CH_3Hg(I)$ available for bioaccumulation. These goals can be met by using bacterial activities because bacteria that are resistant to high concentrations of Hg(II) and $CH_3Hg(I)$ produce enzymes that reduce Hg(II) to the volatile Hg^0 and degrade $CH_3Hg(I)$, respectively (Silver & Walderhaug 1992). The project described here evaluates the use of microbial activities to remove $CH_3Hg(I)$ from contaminated waters.

The study site is a mercury-contaminated freshwater pond, Reality Lake (RL), in the vicinity of Oak Ridge, Tennessee. The source of contamination is a nuclear weapons plant where elemental mercury was used from the 1950s to 1970s to enrich lithium isotopes for the production of hydrogen bombs. Wastewater draining from the plant contains mercury concentrations at µg/L levels and, as a result, fish in the creek downstream from RL occasionally contain $CH_3Hg(I)$ at concentrations above the regulated level of 1 mg/L (Turner et al. 1985).

Research conducted during the last few years has shown that mercury is removed from RL by volatilization as elemental mercury (Hg^0) to the air (Barkay et al. 1991). Experiments indicated that the native microbial community contributed to the reduction of Hg(II) to the volatile Hg^0 (Barkay et al. 1991) and to the degradation of $CH_3Hg(I)$ (Turner et al. 1989). Based on these findings, we have proposed that the concentration of $CH_3Hg(I)$ could be reduced by stimulating microbial activities according to the scheme:

$$\uparrow Hg^0 \leftarrow Hg(II) \leftrightarrow CH_3Hg(I)$$

where the concentration of $CH_3Hg(I)$ is affected directly by degradation (demethylation) and is indirectly reduced by the removal of Hg(II), the substrate for the methylation reaction.

We proposed that the rate of mercury transformations could potentially be accelerated by: (1) adding microbes that reduce Hg(II) to Hg^0 or that degrade $CH_3Hg(I)$, and (2) stimulation of indigenous Hg(II)-reducing and $CH_3Hg(I)$- degrading populations by the addition of growth and energy substrates. We have established a three-tiered system to test these treatments. First, acceleration of the desired reactions is achieved in shake flasks. Second, treatments are applied to microcosms that simulate the cycling of mercury in RL where successful manipulations result in an increase in gaseous mercury (Hg^0) and a decrease in $CH_3Hg(I)$. Finally, treatments that are shown to affect speciation in the microcosms will be applied to enclosures placed in the contaminated pond (Turner et al. 1992). Here we report preliminary results describing the effect of Hg(II)-reducing bacteria on the speciation of mercury in shake flasks and microcosm experiments.

MATERIALS AND METHODS

Isolation of Active Microorganisms and Shake Flask Experiments. RL water and sediment samples were spread on bacterial growth medium containing 10 mg/L Hg(II) (as $HgCl_2$). Strains with the highest tolerance level to Hg(II), as determined by dose response curves, were selected for further testing for their ability to reduce Hg(II) in RL water. These strains were characterized by API strips, and Biolog™ analyses Strain KT17 was characterized as *Klebsiella pneumoniae* (Biolog™ similarity coefficient 0.96) strain KT20 as *Aeromonas schubertii* DNA group 12 (Biolog™ - 0.76) and strain KT23 as *Pseudomonas alcaligenes* (API20 very good ID).

Cells for removal of total mercury (Hg_T) assays were grown to late log phase in rich medium, rinsed twice with RL water and added to Erlenmeyer flasks containing 50 mL RL water to a final concentration of 10^5 cells/mL. Duplicate flasks were set up per each tested strain. Inorganic mercury was added at 1 µg/L, flasks were incubated at room temperature for 24 hr and samples were periodically collected for analysis of remaining Hg_T.

Microcosms Simulating RL. Intact sediment cores (7 cm in depth; 0.5 kg) and overlying water (1.5 L) were collected in 3-L cylindrical jars at the inlet to RL and driven to Gulf Breeze, Florida, within 12 hr of sampling. Each experiment consisted of 4 replicate microcosms. A detailed description of the microcosm will be published elsewhere (Saouter et al. in preparation). Microcosms were incubated at room temperature with a constant input of site water at a turnover time of 8 hr for a period of 3 weeks. The outlet water and the headspace were sampled periodically. Sediments were analyzed at the end of microcosm experiments. Analyses included Hg_T dissolved Hg_T (Hg_{dis}), total $CH_3Hg(I)$ (CH_3Hg_T), and dissolved $CH_3Hg(I)$ (CH_3Hg_{dis}) gaseous mercury (Hg_{gas}) in water leaving the microcosms and Hg^0 in the headspace. These same mercury species were monitored in samples that were obtained in the field on the day of sampling.

Mercury Analysis. Mercury was analyzed using a Brooks Rand Ltd. (Seattle, Washington) cold vapor atomic fluorescence detector by the methods of Gill and Fitzgerald (1987) for Hg_T, and Bloom (1989) for $CH_3Hg(I)$ species.

RESULTS AND DISCUSSION

Removal of Hg_T in Flask Experiments. The addition of bacterial strains to RL water resulted in an accelerated rate of Hg_T removal (Table 1). When available, results of two independent experiments are presented (for RL water and strain KT17). Nonbiological loss of Hg_T from sterile water was apparent, yet water containing indigenous organisms removed Hg_T 2 to 3 times faster as compared to sterile samples. Decline in Hg_T was further accelerated in inoculated samples. The differences between these treatments were statistically significant at $p \leq 0.001$ (ANOVA). The magnitude of Hg_T removal by the added strains varied from slight increase over background (unsterile RL water) for strain KT20 to more than a two-fold increase for strain KT23. In addition, results varied between repeat experiments as is exemplified here for strain KT17 and the unsupplemented control (Table 1).

Although different in magnitude, trends were reproducible. For example, we have consistently observed an increase in evolution of Hg^0

TABLE 1. Hg_T removal from shake flasks containing RL water.

Strains added	Cells/mL	Hg_T loss[(a)] (pg/mL/hr)
Sterile RL water		10.84
RL water[(b)]		31.10
RL water[(b)]		20.42
KT17[(c),(d)]	10^5	41.25
KT17[(c),(d)]	10^5	46.80
KT20[(c)]	10^5	27.60
KT23[(c)]	10^5	69.60

(a) Means of two replicate flasks are presented. Coefficients of variation were 3 to 8%.
(b) RL water containing indigenous flora. Results of two independent experiments are reported.
(c) Cells added to RL water containing indigenous flora.
(d) Results of two independent experiments are reported.

from biologically active RL water as compared to sterile controls (Barkay et al. 1991). Based on these observations, we concluded that amending RL water with bacteria that reduce Hg(II) to volatile Hg^0 resulted in accelerated rates of Hg_T removal and that such amendments could affect the speciation of mercury in the pond.

Microcosms Tests of Remedial Treatments. We previously used mass balance analyses (Watras et al. 1993), accounting for pools and fluxes of mercury species that are dominant in RL, to demonstrate that our microcosm simulated the pond (Saouter et al. manuscript in preparation). Two microcosm replicates were supplemented with strain KT20 (to a concentration of 10^5 cells/mL) after 12 days of equilibrated operation of RL microcosms. The remaining two microcosms served as uninoculated controls. The addition of cells resulted in a 4-fold increase in the amount of Hg^0 that was produced in the microcosm (Figure 1). Because the water column of the microcosms was continuously replaced and the cells were added in a batch (i.e., cells were washed out), a second addition of cells took place 24 hr after the first application. This second addition again resulted in a spike of Hg^0 leaving the microcosm.

The impressive increase in flux of Hg^0 did not affect concentrations of Hg_T or Hg_{dis} in the microcosms (data not shown), probably due to the

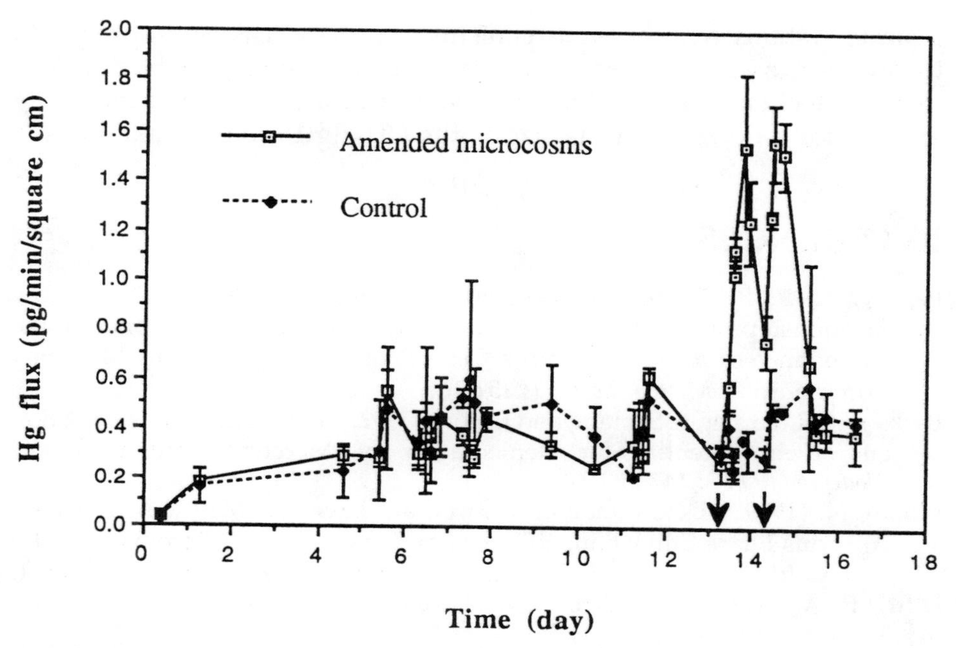

FIGURE 1. Flux of Hg^0 from the surface of microcosms amended with an active Hg(II)-reducing bacterium and from control microcosms. Arrows indicate additions of strain KT20. Bars represent 1 std. deviation of results obtained in 2 replicate microcosms.

continuous input of Hg(II) and the transiency of the added bacterium in the microcosms. Furthermore, a significant decrease in CH_3Hg(I) is not expected under these conditions, as this pool would be affected only as a result of a decrease in concentration of Hg(II), the substrate for the methylation reaction. Thus, maintenance of active Hg(II) reducing populations in the microcosm (and the pond) is essential if a persistent effect on CH_3Hg(I) concentration is to be achieved. Stable maintenance of such populations may be obtained by colonization and growth of the active bacteria in the microcosms or by a continuous input to the microcosms. We are currently developing the technologies that will allow us to maintain a constant number of active cells in the microcosms by a continuous addition with inflow water.

In summary, initial results show that active bacteria that were originally isolated from RL affected the speciation and concentration of mercury in flask incubations and in microcosms that simulate the cycling of mercury in the pond. These results indicate that treating contaminated aquatic systems with active organisms may be a feasible approach to in situ mercury remediation. The problems, it seems, are in the development of technologies to allow persistence of active populations long enough for a decrease in the size of the $CH_3Hg(I)$ pool to take place.

REFERENCES

Barkay, T., R. R. Turner, A. VandenBrook, and C. Liebert. 1991. "The Relationships of Hg(II) Volatilization from a Freshwater Pond to the Abundance of *mer* Genes in the Gene Pool of the Indigenous Microbial Community." *Microb. Ecol. 21*: 151-161.

Barkay, T., R. Turner, E. Saouter, and J. Horn. 1992. "Mercury Biotransformation and their Potential for Remediation of Mercury Contamination." *Biodegradation 3*: 147-159.

Bloom, N. 1989. "Determination of Picogram Levels of Methyl-Mercury by Aqueous Phase Ethylation, Followed by Cryogenic Gas Chromatography with Cold Vapour Atomic Fluorescence." *Can. J. Fish. Aquat. Sci. 46*: 1131.

D'Itri, P. A., and F. M. D'Itri. 1978. "Mercury Contamination: A Human Tragedy." *Environ. manage. 2*: 3.

Gill, G. A. & W. F. Fitzgerald. 1987. "Picomolar Mercury Measurements in Seawater and Other Materials Using Stannous Chloride Reduction and Two-Stage Gold Amalgamation with Gas Phase Detection." *Mar. Chem. 20*: 227-243.

Silver, S., and M. Walderhaug. 1992. "Gene Regulation of Plasmid- and Chromosome-Determined Inorganic Ion Transport in Bacteria." *Microbiol Rev. 56*: 195-228.

Turner, R. R., A. J. VandenBrook, T. Barkay, and J. W. Elwood. 1989. "Volatilization, Methylation, and Demethylation of Mercury in a Mercury-Contaminated Stream." In J. P. Vernet (Ed.), *Proc. Intl. Conf. on Heavy Metals in the Environment*, pp. 353-356. CEP Consultants Ltd., Edinburgh, UK.

Turner, R. R., C. R. Olsen, and W. J. Wilcox. 1985. "Environmental Fate of Hg and ^{137}Cs Discharges from Oak Ridge Facilities." In *Proc. 18th Conf. Trace Substances in Environmental Health*, June 4 - 7, 1984, Columbia, MO.

Turner, R. R., G. R. Southworth, N. A. Bloom, and M. A. Bogle. 1992. "Availability of Sediment-Bound Mercury for Methylation and Bioaccumulation in a Mercury-Contaminated Aquatic System: A Corral Study." In *International Conference on Mercury as a Global Pollutant*, May, 31 - June 4, 1992, Monterey, CA.

Watras, C. J., N. S. Bloom, R. J. M. Hudson, S. A. Gherini, J. G. Wiener, W. F. Fitzgerald and D. B. Porcella. 1993. "Sources and fates of mercury and methylmercury in Wisconsin lakes." In C. Watras and J. Huckabee (Eds.), *Mercury as a Global Pollutant: Toward Integration and Synthesis*. Lewis Publishers, Boca Raton, FL (in press).

BIOREMEDIATION OF SELENIUM OXIDES IN SUBSURFACE AGRICULTURAL DRAINAGE WATER

J. L. Kipps

INTRODUCTION

Inadequate subsurface drainage coupled with increasing salt accumulations on the west side of California's San Joaquin Valley threatens the productivity of more than 4,000 km^2 of irrigated farmland. In the mid-1980s, the disposal of west side agricultural drainage water into the evaporation ponds of Kesterson National Wildlife Refuge resulted in an unprecedented occurrence of waterfowl hatchling deformity and death. The cause was linked to selenium leached from west side soils as selenate [Se(+6)] and its subsequent bioaccumulation. Selenate concentrations in west side drainage water vary from less than 100 to more than 1,000 µg/L. Target drainage treatment levels for selenium are 3 µg/L for evaporation pond disposal and 10 µg/L for discharge into the San Joaquin River (Hanna et al. 1990).

SELENIUM OXIDE BIOREMEDIATION OF AGRICULTURAL DRAINAGE WATER

The development and field testing of selenium-removal agricultural drainage treatment schemes during the last decade revealed that selenium oxide bioremediation requires a combination of denitrification and anaerobic bioreduction of selenate to colloidal elemental selenium, which accumulates in reactor biomass and/or is filtered from reactor effluent. Sufficient carbon (e.g., methanol) is added to the process stream to sustain anaerobic conditions and complete the denitrification process (Hanna et al. 1990).

ADAMS AVENUE BIOLOGICAL SELENIUM-REMOVAL TEST PROGRAM

Program Description. The California Department of Water Resources is sponsoring the Engineering Research Institute at California State University, Fresno, to perform research on pilot-scale, biological selenium-removal processes at the newly constructed Adams Avenue Agricultural Drainage Research Center located about 70 km west of Fresno, California. Additional assistance is provided by the U.S. Bureau of Reclamation and Westlands Water District. This multiyear test program will evaluate the selenium-removal efficiency of one upflow anaerobic sludge blanket reactor (UASBR) and two fluidized bed reactors (FBR) operating at various flowrates. Field operation will emphasize optimizing critical process parameters that require site-specific testing (e.g., reactor performance and wastestream characterization). Once specified, these parameters may be employed in the design of future biological selenium-removal demonstration facilities.

The test program is nearing the completion of Phase I, which focuses on UASBR startup and initial operation. Phase II includes continued operation of the UASBR and startup of two FBRs — one in series, the other in parallel, with the UASBR. Phase III involves continued operation of the UASBR-FBR and FBR test trains and startup of two slow sand filters — one in series with the UASBR-FBR test train and the other in series with the UASBR (Owens 1992).

UASBR Startup. One of the main features of the UASBR process is the gradual formation of granular sludge with a high specific activity and superior settling properties (Lettinga et al. 1980). Granular sludge that forms in fermentative UASBRs has an internal structure dominated by filaments of *Methanothrix* bacteria that form a skeletal matrix to which other bacteria attach (Morgan et al. 1990). The desired microbial processes within the Adams Avenue UASBR are characterized by denitrification and anaerobic selenium oxide bioreduction. To eliminate several months of UASBR startup time dedicated to in situ granular sludge formation, arrangements were made to load the field UASBR with granular sludge from a dormant fermentative UASBR at a bakery wastewater treatment plant in North Kansas City, Missouri. A sample of this granular sludge was analyzed with the following results: average granule size is 1.85 mm, granule settling velocity ranges from 1 to 3 cm/s, and total volatile suspended solids content is 98% (Owens 1992, 1993).

To reduce calcium precipitation during denitrification, initial plans for UASBR operation included feedwater acidification. Four months prior to UASBR startup, two 1-L-capacity UASBRs were seeded with 150-mL granular sludge each. Both reactors received drainage water containing 200 mg/L methanol and 50 mg/L potassium phosphate. Flowrate was set at 1 mL/min, which provided for a hydraulic retention time of 18 hours. The sludge granules were suspended by a recycle flow of 3.785 L/min. One reactor received feedwater dosed with hydrochloric acid to pH 6.9. Within 4 weeks of continuous flow, both reactors demonstrated significant selenium removal, as indicated by the difference between total unfiltered influent selenium concentration and total unfiltered and filtered (0.22-μm filtration) effluent selenium concentrations. After 100 days of continuous operation, percent selenium removals were 79.5% unfiltered and 83.4% filtered for the reactor receiving raw drainage water, and 71.6% unfiltered and 86.8% filtered for the reactor receiving acidified drainage water. Forty days after startup, the granules exposed to the acidified water completely disintegrated into a loose flocculent sludge. Despite the disintegration of the sludge granules, the reactor receiving acidified feed continued to denitrify and reduce selenium, though periodic flotation of the sludge mass was a problem. As a result, plans for acidifying the feedwater to the field UASBR were suspended (Owens 1992).

In September 1992, the 11.4-m^3-capacity UASBR was loaded to capacity with bakery reactor granular sludge and supernatant. The recycle flowrate was set at 303 L/min for sludge bed expansion. After a 4-week period during which increasing quantities of agricultural drainage water were gradually introduced, a continuous flow regime was established with an influent flowrate of 7.57 L/min, which provided for a hydraulic retention time of 25 hours. The nutrient-dosage regime during startup consisted of adding 200 mg/L methanol and 50 mg/L potassium phosphate. Two weeks later, the reactor achieved average selenium removals of 69% unfiltered and 88% filtered. These results demonstrate that the fermentative UASBR granular sludge quickly acquires denitrifying and selenium-reducing capabilities (Owens 1992).

Phase I Results. Figure 1 shows the influence of ambient temperature on UASBR selenium-removal performance for an 18-week period during which influent and recycle flowrates were held constant at 7.57 L/min and 303 L/min, respectively. Nutrient dosage remained unchanged from startup operations. Reactor performance is expressed as percent selenium removal, using values obtained for total unfiltered influent and filtered effluent selenium. Influent selenium concentration ranges from 457 to 592 μg/L. Influent pH averages 7.6 and drops only slightly during

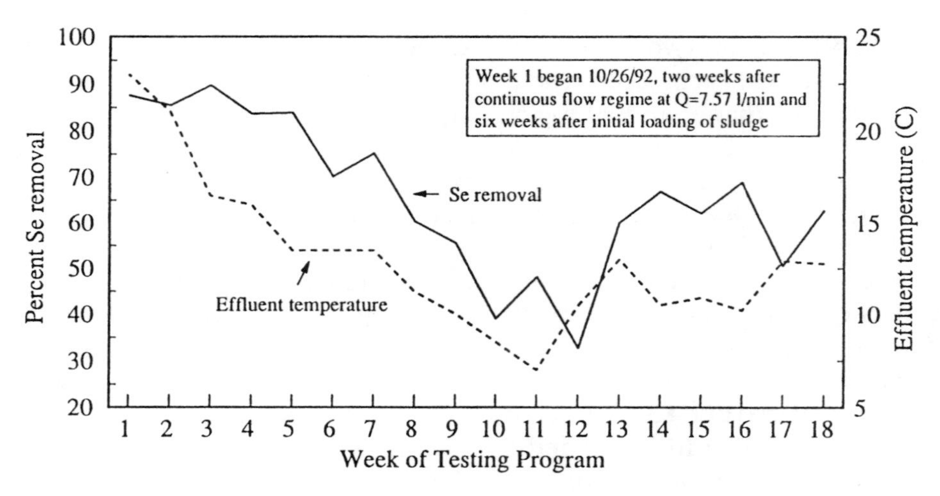

FIGURE 1. UASBR selenium-removal performance and reactor effluent temperature during Phase I.

processing. The reactor's dissolved oxygen content has remained steady at about 0.4 mg/L. Initially, selenium-removal exceeded 80% but dropped to 35% during the coldest weeks of winter. Reactor performance is improving with increasing ambient temperatures (Owens 1993).

In mid-February, an analysis of reactor sludge for volatile suspended solids revealed no change from the value obtained previously for the original sludge. By March, no significant change had occurred to the general appearance of the sludge granules, despite the steady low dissolved oxygen content. A qualitative analysis of reactor gas for methane proved positive. This finding suggests that sludge granules maintain anaerobic microenvironments containing active methanogenic bacteria despite the presence of low dissolved oxygen levels (Owens 1993).

Upcoming Research Activities. Continued UASBR operation will focus on (1) assessing the reactor's performance at various influent flowrates; (2) optimizing operating conditions for selenium-removal performance; and (3) determining the extent to which inorganic solids, especially calcium salts, accumulate in reactor biomass, and how they affect reactor performance. Other key parameters to be monitored include the production and composition of biological reactor gas and the determination of the extent to which selenium and other substances of concern accumulate in reactor biomass.

REFERENCES

Hanna, G. P., Jr., J. L. Kipps, and L. P. Owens. 1990. *Memorandum Report: Agricultural Drainage Treatment Technology Review*. Submitted to San Joaquin Valley Drainage Program, Sacramento, CA.

Lettinga, G., A. F. M. van Velsen, S. W. Hobma, W. de Zeeuw, and A. Klapwijk. 1980. "Use of the Upflow Sludge Blanket (USB) Reactor Concept for Biological Wastewater Treatment, Especially for Anaerobic Treatment." *Biotechnology and Bioengineering* 22:699-734.

Morgan, J. W., L. M. Evison, and C. F. Forster. 1990. "The Internal Architecture of Anaerobic Sludge Granules." *Journal of Chemical Technology and Biotechnology* 50:211-226.

Owens, L. P. 1992. *Adams Avenue Agricultural Drainage Research Center — Status Report, November 1992*. Submitted to California Department of Water Resources, Fresno, CA.

Owens, L. P. 1993. *Adams Avenue Agricultural Drainage Research Center — Status Report, March 1993*. Submitted to California Department of Water Resources, Fresno, CA.

SOIL-GAS AND GROUNDWATER BIOREMEDIATION IN KARST USING IN SITU METHODS

R. E. Moon, E. Nyer, and K. Chellman

INTRODUCTION

A site investigation and in situ bioremediation were conducted at a chemical manufacturing facility to locate and to remove a prominent source of odor, metal corrosion, and groundwater contaminants; 3 acres of a 20-acre facility were highly contaminated. Soil-gas samples contained highly elevated H_2S concentrations and little or no oxygen (<5%); groundwater samples contained elevated concentrations (>10 mg/L) of aromatic solvents. Although H_2S odors had been evident for 10 to 15 years before the investigation, complaints of severe metal corrosion and increased H_2S odor prompted a subsurface investigation. Water and air samples obtained from groundwater monitor wells and soil-gas probes respectively, indicated that the hydrogen sulfide source was generated from a contaminant plume.

The project strategy proposed a two-phased approach: (1) an odor control program and (2) a groundwater remediation program. Each phase was divided into multiple tasks. The odor control program consisted of three tasks: (1) soil-gas sampling, (2) soil vacuum extraction system (SVE) pilot studies, and (3) design and construction of a full-scale SVE system. The odor control strategy proposed the rapid exchange of vapors in the subsurface with ambient air to promote aerobic conditions (Hinchee 1989). The groundwater remediation program was conducted in four tasks: (1) groundwater sampling, (2) bacterial evaluation, (3) groundwater modeling, and (4) design and construction of the remediation system. The investigation and pilot study results provided the basis to design the soil-gas and in situ groundwater bioremediation treatment programs.

SITE DESCRIPTION

Contaminant Origin. During initial facility construction in the 1960s and expansion in the 1970s, subsurface pipelines transferred production wastes to either a waste-water treatment facility or a solvent recycling facility. During this period, subsurface pipelines deteriorated and released liquid wastes into the subsurface. Accidental surface spills and solvent delivery system failures also may have contributed chemicals to the aquifer.

Site Geology/Hydrogeology. The site geology is characterized as karst. All site soil borings reported consolidated limestone from land surface to approximately 6 m below land surface (BLS). Between 6 m and 9 m BLS, borings logs and sample corings exhibited cavernous areas and highly porous features.

The groundwater flow direction and aquifer characteristics were determined by simultaneous water-table elevation measurements and pumping tests, respectively. Water-table elevation fluctuations between 0.5 m and 0.75 m corresponded to tide responses. A freshwater lens (ranging from 2.1 to 4.6 m thick) was present beneath the facility. This condition required the groundwater remediation program to protect the freshwater lens from saltwater intrusion.

SITE INVESTIGATION SUMMARY

Six field and modeling activities provided the basis to design the in situ bioremediation system. The conclusions of each investigation are summarized below:

- The soil-gas investigation located the subsurface odor source and supported the proposed groundwater recovery well locations.
- The SVE pilot studies demonstrated that ambient air could be transferred to the subsurface via air inlet wells. The studies showed that an SVE well could create a 15-m radius of influence and the optimal depth for the screened interval of an SVE well was 3 to 4.6 m BLS.
- The full-scale SVE design captured soil-gases from the entire subsurface that produced odor. After 2 years of operation, the H_2S concentrations measurements in the subsurface had

decreased significantly below detection limits (BDL); oxygen concentrations increased to near ambient levels.

- The groundwater investigation defined the horizontal extent of the contaminant plume and identified the specific constituents that constituted the odor source.
- The bacterial evaluation confirmed the presence of heterotrophic aerobic bacteria in limestone core samples and the ability of in situ bacteria to metabolize groundwater contaminants. Bacterial enumerations confirmed the presence of extreme anoxic conditions within the contaminant plume and aerobic conditions in unaffected areas. The nutrient adsorption study confirmed that nutrients would not interact with the limestone and limit their availability for in situ bioremediation. Water quality analyses confirmed the need to supplement groundwater with additional nutrients.
- The groundwater modeling program determined the appropriate pumping rate for groundwater recovery wells necessary to avoid saltwater intrusion near recovery wells. The groundwater model was also used to select the recovery well locations.

GROUNDWATER REMEDIAL DESIGN AND CONSTRUCTION

The in situ bioremediation system was designed to fulfill three tasks: (1) recover groundwater from the contaminant plume; (2) remove groundwater contaminants via aeration and microbial oxidation; and (3) introduce oxygen, nutrients, and aerobic bacteria into the aquifer to enhance in situ degradation of contaminants.

The in situ groundwater treatment system consisted on 13 groundwater recovery wells, the aeration unit, and 10 injection wells. The injection wells were designed to allow the partially treated groundwater to cascade into an open boring. This discharged method minimized bacterial fouling and limited saltwater and freshwater mixing.

The main operational unit of the treatment system was a groundwater aeration unit. The design was based on trickling filter technology. Packed media (1.1 m^2/m^3) were placed in a square container (3.2 m × 3.2 m) constructed of concrete. The media were 1.8 m deep and provided surface area for the aerobic bacteria to attach and grow. Oxygen was transferred into the groundwater as it moved through the media. Nutrients were added to the groundwater before the aeration unit. The treatment unit introduced microbes into the partially treated groundwater; aerobic bacteria, growing on the packed media, were dislodged from the force of the cascading water and transferred into the aquifer via the injection wells. Based on optimum flowrates obtained from groundwater modeling, the aeration unit was designed for a flowrate of 492 L/min.

In Situ Groundwater Bioremediation Results. Volatile organic compound (VOC) analyses of groundwater samples showed a significant reduction among individual constituents during the first 8 months of operation. Toluene concentrations initially ranged from 10 mg/L to 10,000 mg/L in all groundwater monitoring wells. Following the removal of 5 pore volume (22.5 million gallons), the concentration range declined; toluene concentrations in groundwater samples ranged from less than 10 mg/L to 1,000 mg/L.

The aeration unit has resulted in the near-complete removal of organic constituents from the groundwater. A comparison of VOC removal efficiency between influent and effluent samples from the aeration unit showed an increase from an initial removal percentage of 93% to 99% after 2 months; the dissolved oxygen content in the aeration unit effluent samples increased from 2.3 mg/L to 4.5 mg/L. Within several weeks of operation, a thick bacterial mat formed on the aeration unit's packed media. Although the performance of these bacteria to degrade the groundwater contaminants was not measured, it was apparent that a large population of aerobic bacteria was available for transport and introduction into the aquifer.

Among the recovery wells, individual chemical analyses showed VOC reductions ranging from 94% (toluene) to 100% [methyl tertbutyl ether (MTBE), ethylbenzene, and xylene] when the average VOC concentrations in water samples were compared.

A comparison between the total VOC concentrations and the total reduced sulfur concentrations recorded during the project showed a sustained decline in the concentration of reduced sulfur compounds following the installation of the SVE. In addition, the total VOC concentration diminished prior to and after the groundwater recovery system was installed.

Salinity measurements and total dissolved solids (TDS) content were obtained from all groundwater monitoring wells as part of a routine sampling program. Following 8 months of operation, there was no apparent mixing of the freshwater lens with salt water.

REFERENCE

Hinchee, R. E. 1989. "Enhanced Biodegradation Through Soil Venting." *Proceedings of the Workshop on Soil Vacuum Extraction*, Robert S. Kerr Environmental Research Laboratory, Ada, OK, April 27-28.

INTRASPECIFIC TRANSFER OF ORGANIC XENOBIOTIC CATABOLIC PATHWAYS TO CONSTRUCT BACTERIA OF ENVIRONMENTAL INTEREST, ADAPTED FOR ORGANIC XENOBIOTIC DEGRADATION IN PRESENCE OF HEAVY METALS

D. Springael, L. Diels, J. van Thor, A. Ryngaert, J. R. Parsons, L. C. M. Commandeur, and M. Mergeay

INTRODUCTION

Presence of copollutants such as heavy metals in organic xenobiotic contaminated environments occurs frequently and may have a negative effect on the survival and degradative capacity of introduced microorganisms specialized to degrade the organic contaminant of interest (Kovalick 1991, Said & Lewis 1991, Springael et al. 1993). Therefore, engineering of organic xenobiotic-degrading bacteria through introduction of heavy metal-resistant genes or of heavy metal-resistant bacteria through introduction of catabolic genetic information can be a useful tool. In a previous paper, the construction was reported of heavy metal-resistant haloaromatic-degrading *Alcaligenes eutrophus* strains by simple mating procedures between strains carrying plasmid-encoded heavy metal resistance determinants and haloaromatic degradation (Springael et al. 1993a). In this paper, we describe the transfer of polychlorinated biphenyl (PCB) catabolic genes, which are chromosomally bound and thus difficult to transfer between bacteria, from PCB degraders to heavy metal-resistant bacteria and the genetic events involved.

MATERIALS AND METHODS

Strains and Plasmids. The *Alcaligenes eutrophus* A5 strain is able to use biphenyl (BP) and 4-chlorobiphenyl (4CBP) as sole sources of carbon and energy (Shields et al. 1985). The genetic information for BP/4CBP degradation is carried on a 60-kb large transposable element, Tn*4371*, present on the chromosome (Springael et al. 1993b). *A. eutrophus* A5 carries the 51-kb IncP1 plasmid pSS50. A5.79 is a derivative of A5 but carries the 60-kb IncP1 plasmid RP4 instead of pSS50. *Alcaligenes denitrificans* JB1 is able to use different aromatic xenobiotic compounds as a carbon source such as BP, 4CBP, 3-chlorobiphenyl (3CBP), naphthalene, and *m*- and *p*-toluic acid and can cometabolize several higher chlorinated PCBs and dioxins (Parsons et al. 1988). The genetic information for the degradation of these compounds is carried on the chromosome. AE1143 is a JB1 derivative strain carrying the RP4::Mu3A plasmid. RP4::Mu3A is a MiniMu derivative of phage Mu and behaves as a very effective transposable element which is able to pick up genomic DNA (formation of R prime plasmids) to transfer it to other bacteria (Lejeune et al. 1983). *A. eutrophus* CH34 contains the megaplasmids pMOL28 specifying nickel, mercury, chromate, cobalt, and thallium resistance and pMOL30 specifying zinc, cadmium, cobalt, mercury, and copper resistance (Mergeay et al. 1985).

Matings. Matings were done as described by Lejeune et al. (1983).

Plasmid Extractions. Plasmid extractions were done as described by Kado and Liu (1981).

RESULTS

Transfer of the PCB Catabolic Pathway of* A. eutrophus *A5 to the Heavy Metal-Resistant Bacterium* A. eutrophus *CH34: Transposition of the PCB Catabolic Genes on a Transferable Plasmid. Both *A. eutrophus* A5 and A5.79 were mated with *A. eutrophus* CH34, and PCB-degrading metal-resistant transconjugants were selected on Tris minimal medium (Mergeay et al. 1985) containing BP as a carbon source and nickel. With both donors, the BP/PCB catabolic marker was transferred to CH34 with a frequency of 10^{-6} per recipient cell. Transconjugant CH34 strains carried an enlarged pSS50 or RP4 plasmid containing the BP/4CBP degradative transposon Tn*4371*. The strains were able to grow on BP and 4CBP in the presence of various heavy metals whose resistance was specified by plasmids pMOL28 and pMOL30. One of these transconjugant strains, AE707, was shown previously to degrade several PCB isomers in the presence of nickel or zinc (Springael et al. 1993a).

Transfer of the PCB Catabolic Pathway of* A. denitrificans *JB1 to the Heavy Metal-Resistant Bacterium* A. eutrophus *CH34: Formation and Transfer of R Prime Plasmids Carrying Chromosomal-Encoded PCB Catabolic Genes. *A. denitrificans* JB1 derivative AE1143 was mated with *A. eutrophus* CH34, and PCB-degrading metal-resistant transconjugants were selected on Tris minimal medium containing BP as a carbon source and nickel. The BP/PCB degradative pathway of JB1 was transferred to CH34 by means of RP4::Mu3A at a frequency of 10^{-8} per recipient cells. Plasmid analysis of transconjugants revealed the presence of RP4::Mu3A R prime plasmids carrying DNA inserts of varying length (50 to 100 kb). The R prime plasmids showed a significant degree of instability when grown on unselective media, whereas integration in the chromosome was observed when the strains were grown on selective medium with BP. PstI restriction enzyme analysis of three R prime plasmids demonstrated the presence of identical extra fragments in the 3 plasmids. The constructed strains were able to grow on BP and 4CBP in the presence of various heavy metals.

CONCLUSIONS

Two genetic methods are presented to transfer a desired genetic trait, i.e. organic xenobiotic degradation to another bacterium of interest, i.e., heavy metal-resistant bacteria. The first one requires the presence of the trait on a transposable element. However, the second method can be used to transfer any desired phenotype between bacteria in which RP4::Mu3A is able to replicate and transfer. Apart from PCB degradation, we were able to transfer another aromatic degradative pathway, i.e., metabolism of *m*-hydroxybenzoate from JB1 to CH34. As such, PCB-degrading, heavy metal-resistant bacteria were constructed that can be used to decontaminate organic polluted sites that also contain various heavy metals. Indeed, about 35% of polluted sites in the USA exhibit a combined pollution of organics and inorganics (Kovalick 1991).

REFERENCES

Kado, C. I., and S. T. Liu. 1981. "Rapid procedure for detection and isolation of large and small plasmids." *J. Bacteriol. 145*: 1365-1373.

Kovalick, W. 1991. "Perspectives on health and environmental risks of soil pollution and experiences with innovative remediation technologies." Abstr. 3.3-1.

Abstr. 4th World Congress Chemical Engineering. 16-21 June, 1991, Karlsruhe, Germany.

Lejeune, P., M. Mergeay, F. van Gijseghem, M. Faelen, J. Gerits, and A. Toussaint. 1983. "Chromosome transfer and R-prime formation mediated by plasmid pULB113 (RP4::miniMu) in *Alcaligenes eutrophus* CH34 and *Pseudomonas fluorescens* 6.2." *J. Bacteriol. 155*: 1015-1026.

Mergeay, M., D. Nies, H. G. Schlegel, J. Gerits, P. Charles, and F. van Gijseghem. 1985. "*Alcaligenes eutrophus* CH34 is a facultative chemolithotroph with plasmid bound resistance to heavy metals." *J. Bacteriol. 162*: 328-334.

Parsons, J. R., D. T. H. M. Sijm, A. van Laar, and O. Hutzinger. 1988. "Biodegradation of chlorinated biphenyls and benzoic acids by a *Pseudomonas* strain." *Appl. Microbiol. Biotechnol. 29*: 81-84.

Said, W. A., and D. L. Lewis. 1991. "Quantitative assessment of the effects of metals on microbial degradation of organic chemicals." *Appl. Environ. Microbiol. 57*: 1498-1503.

Shields, M. S., S. W. Hooper, and G. S. Sayler. 1985. "Plasmid-mediated mineralization of 4-chlorobiphenyl." *J. Bacteriol. 163*: 882-889.

Springael D., L. Diels, L. Hooyberghs, S. Kreps, and M. Mergeay. 1993a. "Construction and characterization of heavy metal resistant haloaromatic degrading *Alcaligenes eutrophus* strains." *Appl. Environ. Microbiol. 59*: 334-339.

Springael, D., S. Kreps, and M. Mergeay. 1993b. "Identification of a transposable element, Tn*4371*, carrying biphenyl and 4-chlorobiphenyl degradation genes in *Alcaligenes eutrophus* A5." *J. Bacteriol.* In press.

RELEASE OF METAL-BINDING FLOCCULENTS BY MICROBIAL MATS

S. Rodriguez-Eaton, U. Ekanemesang, and J. Bender

INTRODUCTION

Research performed in our laboratory uses mixed microbial mats designed for specific problems in bioremediation. The mats are composed of cyanobacteria, the dominant group being *Oscillatoria* spp., as well as heterotrophic and ensiling bacteria. The resulting constructed mats are durable, resilient, and tolerant of wide environmental fluctuations. These mats have been used successfully to promote the uptake of heavy metals from contaminated waters, as well as to degrade chlordane, TNT, and petroleum distillates (Bender & Phillips 1994).

Due to the very complex nature of microbial mats, few mechanism(s) responsible for their bioremediation capabilities are known. Some species of benthic cyanobacteria have been shown to produce extracellular secretions with the capability of flocculating and sedimenting clay particles as a survival mechanism (Bar-Or & Shilo 1987). We have observed, upon challenging a mat with heavy metals, the production and secretion of proteins (flocculents) and their accumulation at the bottom of the water column. Here we attempted to determine the correlation between bioflocculent production and metal sequestration by the microbial mats. To accomplish this we examined two technical applications for the metal removal; a column packed with glass wool and capped on both ends with immobilized microbial mat, and a free floating ball with immobilized mat (approximately 1/4 the biomass size of the column).

METHODS

Microbial mats were constructed by integrating metal-tolerant bacteria and cyanobacteria from laboratory stock cultures (Bender et al. 1989).

Two different experiments were designed and performed in duplicate to investigate the correlation between bioflocculent production and metal sequestration. One-way ANOVA was performed to test for significant differences in bioflocculent production between treatments (Steel & Torrie 1960).

Mat-Covered Ball Experiment. In a tray containing 10% Allen and Arnon medium (1955) and 5 g of silage, 4-gram glass wool balls were inoculated with microbial mat. Mat-covered balls matured (all surfaces covered with mat) in 7 to 10 days. The mature balls were individually submerged in glass trays containing 1 L of test metal or control solution. Daily samples for determining metal and bioflocculent concentrations were taken from initiation of the study (day 0).

Two different of metal ion solutions were used. Acid coal mine drainage came from a field site in Alabama, containing 3.5 mg/L Fe and 5.3 mg/L Mn (pH 5.5). Influent iron was likely ferrous (Fe^{+2}) iron that became oxidized to ferric (Fe^{+3}) iron upon contact with the atmosphere. Influent manganese was in solution as Mn^{+2}. Synthetic solutions for laboratory experiments were prepared from $Mn(NO_3)_2$ and $Zn(NO_3)_2$ at concentrations of 36 mg/L Zn and 38 mg/L Mn (pH 4.5). Controls used distilled water.

Column Experiment. Six glass columns, each containing 8 g of glass wool at the bottom and at the top (total glass wool content was 16 g/column), were inoculated with mat. A 1,200-mL water column (10% Allen and Arnon medium) separated the two mats. Water columns were covered with foil to prevent extensions of algal growth. When mats covered the full diameter of top and bottom columns, the experiment was initiated. Daily samples for determining metal and bioflocculent production were taken.

The columns were spiked with concentrated metal solutions so as to minimally increase the total aqueous volume/column. In the first column set, one metal addition was made (final concentrations, mg/L, on day 0: 18 Zn and 19 Mn). In the second set, three additions were made (days 0, 6, and 12) to maintain nearly constant concentrations (mg/L) of 18 Zn and 20 Mn. Therefore, in this latter set, total metal addition was much higher than the first set. Controls received no metal additions. The purpose of the second set was to increase the bioflocculent production by maintaining high metal concentrations in the water column.

Metal and Bioflocculent Determinations. Metal analysis of the samples consisted of the following steps. 6N HNO_3 was added in a 1:10 dilution. Samples were treated by microwave digestion (CEM Corp. model MDS-200) and analyzed by atomic absorption (Varian model Spectra AA-20 BQ, double beam). Bioflocculent was determined with the Alcian Blue Binding Test (Bar-Or & Shilo 1987). Alcian blue 8NGX was dissolved in 0.5 M acetic acid to a final concentration of 1 mg/mL. A 1-mL water column sample, 3.75 mL 0.5 M acetic acid, and 0.25 mL alcian blue dye were mixed and incubated overnight. Samples were centrifuged for 20 minutes at no less than 5,000 rpm. Absorbance (outer diameter, OD) was read at 610 nm. Control OD was determined with 0.5 M acetic acid in 10% Allen and Arnon media.

RESULTS

Metal exit from the water column was achieved for all experimental conditions in both experiments. The experiments were performed under quiescent conditions. Mixing would likely increase the diffusion rate of the bioflocculent and thereby increase the interactions with the metals.

There was no significant difference between duplicates. Tables 1 and 2 show the results of one representative set from each experiment.

Mat-Covered Ball Experiment. In trays containing acid coal mine drainage, the pH increased from 5.3 to 8.4, and Mn and Fe precipitated

TABLE 1. Mat-covered ball experiment.

	Acid Coal Mine Drainage			Synthetic Solution			Control
Day	Metals, Fe	mg/L Mn	Biofloc. (1/OD)	Metals Zn	mg/L Mn	Biofloc. (1/OD)	Biofloc. (1/OD)
0	3.14	5.4	0.73	30.1	37.9	0.71	0.69
1	0.05	1.1	0.95	26.5	34.9	0.72	0.70
2	0.07	0.98	1.16	29.8	32.3	0.77	0.71
3	0.61	0.34	1.27	27.1	34.9	0.77	0.73
4	0.05	0.44	1.37	28.7	34.4	0.91	0.74
5	0.08	0.31	1.41	31.8	41.0	1.05	0.77
6	0.03	0.33	1.45	31.8	33.4	1.16	0.78
7	0.03	0.27	1.56	29.3	33.6	1.16	0.80
8	0.03	0.20	1.70	28.9	32.1	1.22	0.81
9	0.04	0.22	1.85	24.5	32.0	1.25	0.83
10	0.03	0.11	2.04	9.6	6.9	1.45	0.85
11	0.02	0.07	2.17	3.4	1.9	1.59	0.86

One-way ANOVA showed the following significant differences between the two experimental and control treatments: (1) Acid mine drainage produced more bioflocculent than the synthetic solution ($P < 0.01$); (2) Bioflocculent production was greater in acid mine drainage ($P < 0.01$) and synthetic solution ($P < 0.05$) than in the control solution.

TABLE 2. Column experiment.

	Mn/Zn Added Day 0			Mn/Zn Added Day 0, 6, 12			Control
Day	Metals, Fe	mg/L Mn	Biofloc. (1/OD)	Metals Zn	mg/L Mn	Biofloc. (1/OD)	Biofloc. (1/OD)
0	19.3	17.5	0.64	17.5	20.1	0.57	0.77
1	15.9	16.3	0.64	14.9	18.1	0.75	0.77
2	11.9	10.9	0.69	12.1	16.3	0.92	0.75
3	8.6	7.3	0.69	8.6	10.3	1.0	0.74
4	7.9	6.9	0.81	7.5	9.6	1.08	0.74
5	6.3	5.7	0.94	6.3	8.5	1.11	0.73
6	5.2	4.6	1.04	4.2	5.2	1.15	0.81
6*				16.9	19.3		
7	4.6	3.3	1.04	14.5	16.4	1.23	0.92
8	3.8	2.8	1.09	12.9	11.6	1.34	0.93
9	2.1	2.3	1.11	7.5	9.7	1.46	0.94
10	0.9	1.5	1.29	7.0	7.3	1.47	0.92
11	0.5	0.9	1.41	6.3	6.9	1.49	0.89
12	0.4	0.8	1.42	4.9	5.3	1.51	0.91
12*				19.3	21.5		
13	0.3	0.7	1.45	13.5	15.4	1.61	0.91
14	0.3	0.4	1.47	6.9	9.5	1.63	0.89
15	0.3	0.3	1.49	5.3	7.3	1.71	0.88

Bioflocculent production was significantly greater in metal-containing columns versus control columns (day 0 > control, $P < 0.05$; days 0, 6, 12 > control, $P < 0.01$).

on day 1. The 79.6% removal of Mn on the first day may be due in part to a coprecipitation with iron.

Synthetic solutions containing (mg/L) approximately 30 Zn and 38 Mn, (initial pH 4.5) showed a slower metal exit (Table 1). By day 11, 89% of Zn and 95% of Mn were removed. Both experimental conditions showed a higher level of flocculent production as compared to controls and the acid coal drainage showed more bioflocculent than the synthetic solution (Table 1).

Column Experiment. Within 6 days, 73 to 76% of Zn and Mn was removed from all water columns. Both the spiked and non-spiked columns showed continuously increasing bioflocculent production throughout the duration of the experiment (Table 2). Both were significantly different from the controls. However, there was no significant difference between the spiked and non-spiked columns in bioflocculent production.

CONCLUSIONS

Correlations between metal removal and bioflocculent production can be seen in that low production corresponds to slower metal deposit and as the activity increased the metal exit increases (Table 1, synthetic solution). Once stimulated, the bioflocculent production seems to continue even after the metals have been essentially removed (Table 1, acid coal mine drainage). Additionally, spiking the metals did not increase the bioflocculent production (Table 2).

Although there is a clear correlation between metal exit and bioflocculating activity in the synthetic laboratory solution containing only Zn and Mn, the field sample, containing Fe and Mn does not show the same correlations. In the field sample Fe and Mn deposited more rapidly, whereas the bioflocculating activity emerged more slowly. Fe is rapidly oxidized by the photosynthetic oxygen and subsequently deposits as $Fe(OH)_3$. Mn is known to coprecipitate with Fe under the highly toxic and elevated pH conditions, such as those mediated by the mat (Bender & Phillips 1994). Thus these environmental effects were likely the dominant metal-depositing factors in the field and probably masked the effects of the bioflocculent.

These results show that the presence of Mn and/or Zn stimulates the production of bioflocculents by the microbial mats. This production, once induced, is continuous and is not increased by continuous spiking of the metals.

ACKNOWLEDGMENTS

Infrastructure support was provided by U.S. Environmental Protection Agency Grant No. CR81868901. Project support for bioflocculent analysis was provided by the Office of Naval Research N00014-91-J-1823.

REFERENCES

Allen, M. B., and D. I. Arnon. 1955. "Studies on Nitrogen-fixing Blue-green Algae. I. Growth and Nitrogen Fixation by *Anabaena cylindrica* Lemm." *Pl. Physiol., Lancaster 30*:366-372.

Bar-Or, Y., and M. Shilo. 1987. "Characterization of Macromolecular Flocculents Produced by *Phormidium* J-1 and by *Anabaenopsis circularis* PCC 6720." *Appl. Environ. Microbiol. 53*:2226-2230.

Bender, J., E. R. Archibold, V. Ibeanusi, and J. P. Gould. 1989. "Lead Removal from Contaminated Water by a Mixed Microbial Ecosystem." *Water and Science Technology 21*(12):1661.

Bender, J., and P. Phillips. 1994. "Implementation of Microbial Mats for Bioremediation." In J. L. Means and R. E. Hinchee (Eds.), *Emerging Technology for Bioremediation of Metals*. Lewis Publishers, Ann Arbor, MI. pp. 85-98.

Steel, R. G. D., and J. H. Torrie. 1960. *Principles and Procedures of Statistics*. McGraw-Hill Book Co., Inc., New York, NY.

REMEDIATION OF MINING WATER WITH MICROBIAL MATS

Y. Vatcharapijarn, B. Graves, and J. Bender

INTRODUCTION

Microbial mats are coherent, organic-rich, multilayered communities formed by several groups of microbes adhering to substrates in aquatic environments. This coherence arises from the binding of microbial filaments, extracellular slime, or precipitated minerals. Photosynthetic cyanobacteria, which dominate microbial mats, provide nutrient support to the community and several other factors that likely contribute to the metal-sequestering and metal-storage capacity of this consortium.

MATERIALS AND METHODS

Microorganisms were enhanced for metal tolerance by stepwise exposure to increasing metal concentrations (Bender et al. 1989). Metal-tolerant bacteria (mixed species of soil bacteria) and cyanobacteria (*Oscillatoria* spp.) were integrated into microbial mats, which were subsequently used in the metal-sequestering experiments. Procedures for laboratory development of constructed microbial mats followed Bender and Phillips (1994).

Microbial mats were immobilized on glass wool by broadcasting mat pieces on a layer of glass wool saturated with growth medium and maintained under incandescent lamps. Mature immobilized mats were assessed for metal-sequestering capacity in the following three applications: baffled tanks (61 × 18 cm with 4.5 cm depth of glass wool), columns (25 cm × 3.7 cm diameter), and glass wool floaters (5 cm × 5 cm × 2 cm). Metal-contaminated water, passed through the baffled tanks and columns, was measured for metal concentration. Floaters were applied to metal-contaminated solutions, and water columns were routinely sampled over a 12-day period.

Metal Application Sampling and Analysis. Samples (water columns or mats) were hydrolyzed by microwave digestion (CEM Corp. Model MDS-2000). Pre-digestion was followed by a compressed digestion using 6M HNO_3, with a 1:10 dilution (acid:sample), for 30 minutes. Hydrolysates were analyzed by atomic absorption (Varian model, Spectra AA-20 BQ, double beam).

Three baffled tanks were used in series. Effluents from the first tank were applied to the second and third tanks. Six applications of 0.5 L of mixed Zn (16 mg/L) and Mn (15 mg/L) solutions were passed through the tanks. Controls had a glass wool layer without mat. The flowrate through the system was 3.3 mL/min.

Columns packed with immobilized mats were used to treat mixed solutions of Zn and Mn (14 mg/L each metal). Control columns contained glass wool only and glass wool with ensiled grass. Mats were immobilized on glass wool by broadcasting sections of the metal-tolerant mat on glass-wool packed into acrylic columns. In 7 days columns were ready for metal applications; 15 applications were made over a period of 15 days. There was no renutrification of the mats between applications.

Floaters (containing 9 cm^2 of mat on the surface) were applied to 1 L of metal solution containing Zn and Mn (20 and 18 mg/L, respectively). Water columns were analyzed for metal concentration over a 12-day period. To increase the metal-packing capacity, the floaters were renutrified every third day by placing them in nutrient solution and allowing for new cell growth. Controls, containing glass wool floaters (without mats), were cycled through the nutrification medium in a treatment similar to that of the experimental floaters.

RESULTS AND DISCUSSION

The three applications of immobilized mats showed good removal of Zn and Mn, applied in mixed solution. Control wool, without mat, sequestered Zn. However, the effect of the biological material on metal sequestering clearly is demonstrated.

Figure 1 is representative of a serial flow scheme through the three baffled tanks. Effluent concentrations from each of the baffled tanks showed good metal removal. By the final treatment (tank three) both metals showed an average of 98 to 99% removal. Zn typically bonds to glass wool until attachment sites are saturated. This can be seen at tank 1. Mn had little attachment to glass wool.

Columns, packed with immobilized mat, showed good removal for the first 13 flows, then rapidly diminished in flows 14 and 15 (Figure 2).

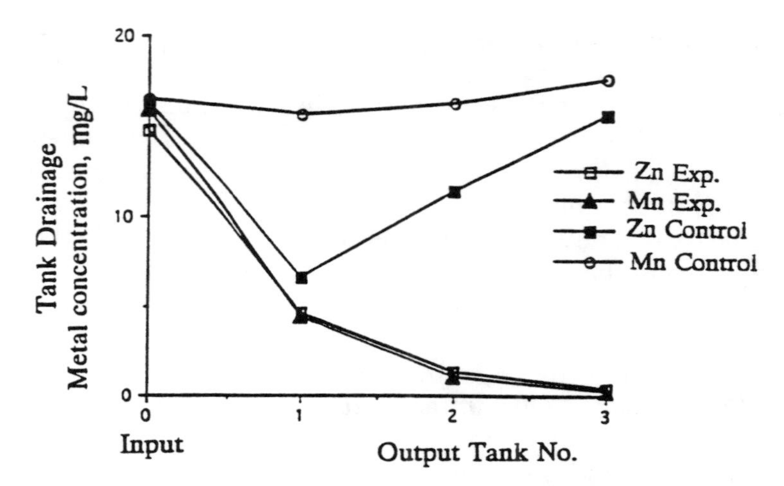

FIGURE 1. Removal of Zn and Mn, in mixed solution, by immobilized mat layered in baffled tanks. Three baffled tanks (causing solutions to flow in a "snaking" movement) were used in series to treat Zn and Mn mixed solution. Effluent from tank 1 was passed through tanks 2 and 3. Controls contained glass wool immobilizing substrate without mat (Bender et al., submitted to *Water, Air and Soil Pollution*).

X-ray microanalysis of several mat locations showed no congruent deposit of the metal with the cells (data not shown). The amorphous metal precipitates along the surface of the mat suggest that the metals were deposited by chemical conditions mediated by the mat, rather than by cell sorption. The decrease in metal removal (flow #13) may be caused by a interruption of photosynthesis due to a darkening of the mat surface by metal deposit. However the floater experiments, described below, indicated that this can be alleviated by renutrifying the mats to allow new surface growth of the cyanobacteria.

Floaters showed consistent metal removal for Zn from treatment days 4 through 12 (Figure 3). Mn uptake increased as the floaters were nutrified and reapplied to the metal solution. Mat analysis showed that 81% of the Zn and 70% of the Mn was deposited in the mat and its associated organic material. The remaining metals were found in the flocculated material at the bottom of the reaction tray and renutrification beakers. Metals associated with the mats were easily removed by harvesting the floaters from the water column.

In all immobilized mats, the microbial cells maintained their viability. Extracellular metal deposit may have contributed to a lack of toxicity to

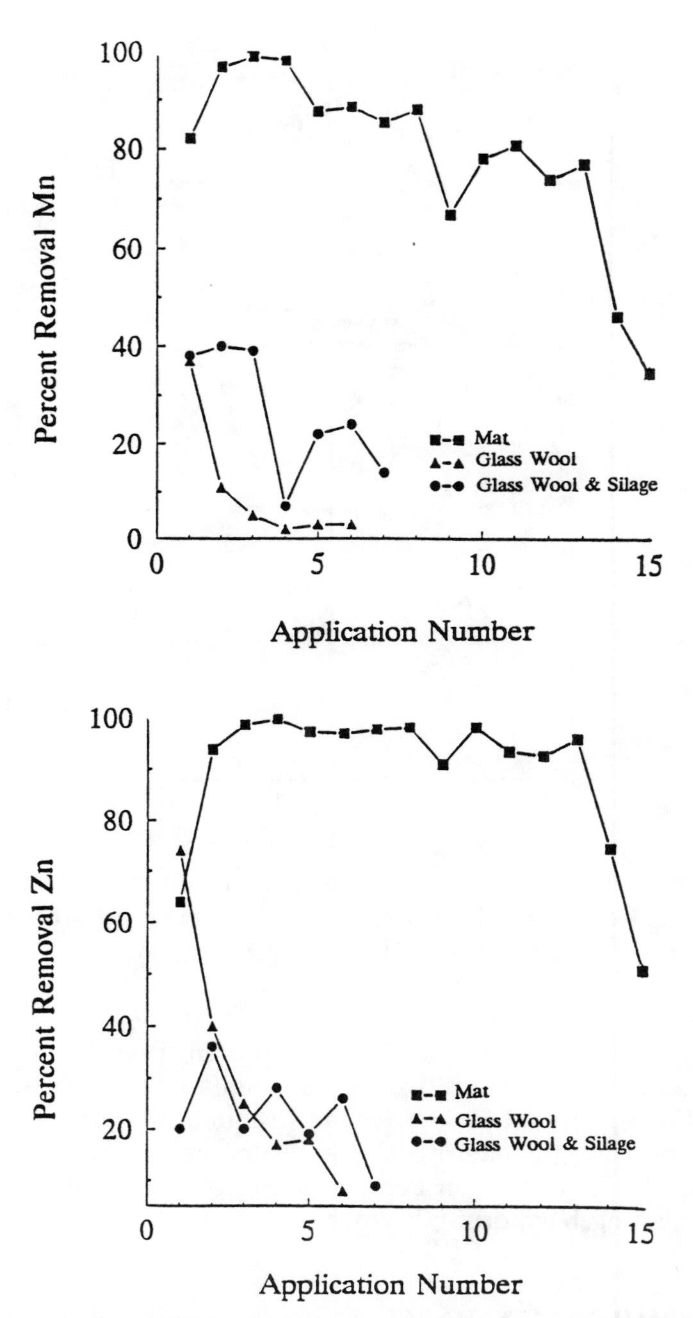

FIGURE 2. Removal of Mn and Zn, in mixed solution, by immobilized mat packed in columns. Mn/Zn solution was applied 15 times to the mat/column. Effluent metal concentrations determined percents removal (Bender et al., submitted to *Water Environment Federation*).

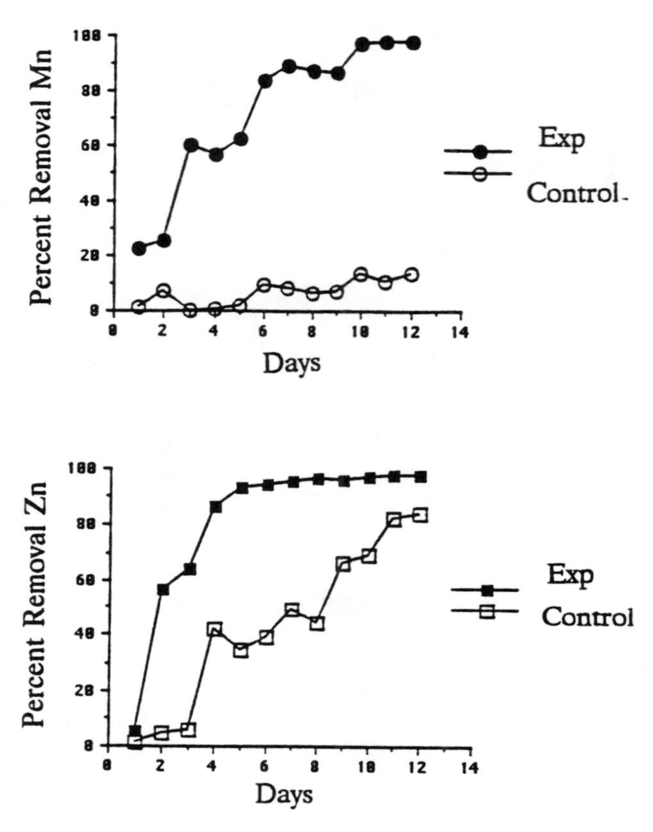

FIGURE 3. Removal of Mn and Zn, in mixed solution, by immobilized mats applied as free-floating units. All floaters (immobilized mats and controls) were renutrified every third day.

the biological system. Photosynthetic oxygen, present near the mat surfaces, and elevated pH and Eh levels, mediated by the mat, favor the precipitation of Zn and Mn oxides in the aqueous environment. Metal removal induced by altering the environmental chemistry may be preferred over cell sorption because it protects the cells and thereby extends the long-term durability of the system.

ACKNOWLEDGMENTS

Infrastructure support was provided by U.S. Environmental Protection Agency Grant No. CR81868901. Project support for the immobilized mat system was provided by the U.S. Bureau of Mines Grant No. G0190028 and U.S. Dept. of Army Grant DACA39-91-K-0004.

REFERENCES

Bender, J., and P. Phillips. 1994. "Implementation of Microbial Mats for Bioremediation." In J. L. Means and R. E. Hinchee (Eds.), *Emerging Technology for Bioremediation of Metals*. Lewis Publishers, Ann Arbor, MI. pp. 85-98.

Bender, J., J. P. Gould, Y. Vatcharapijarn, and J. S. Young. Under review. "Removal of Zinc and Manganese From Contaminated Water with Cyanobacteria Mats." *Water, Air and Soil Pollution*.

Bender, J., J. R. Washington, B. Graves, P. Phillips, and G. Abotsi. In press. "Deposit of Zinc and Manganese in a Aqueous Environment Mediated by Microbial Mats." *Water Environment Federation*.

Bender, J., E. R. Archibold, V. Ibeanusi, and J. Gould. 1989. "Lead Removal from Contaminated Water by a Mixed Microbial System." *Water Science and Technology, 21*(12): 1661-1665.

POLYURETHANE AND ALGINATE-IMMOBILIZED ALGAL BIOMASS FOR THE REMOVAL OF AQUEOUS TOXIC METALS

I. V. Fry and R. J. Mehlhorn

INTRODUCTION

We describe the development of immobilized, processed algal biomass for use as an adsorptive filter in the removal of toxic metals from wastewater. To fabricate an adsorptive filter from processed biomass, several crucial criteria must be met: (1) high metal-binding capacity, (2) long-term stability (both mechanical and chemical), (3) selectivity for metals of concern (with regard to ionic competition), (4) acceptable flow capacity (to handle large volumes in short time frames), and (5) stripping/regeneration (to recycle the adsorptive filter and concentrate the toxic metals to manageable volumes). This report documents experiments with processed algal biomass (*Spirulina platensis* and *Spirulina maxima*) immobilized in either alginate gel or preformed polyurethane foam. The adsorptive characteristics of these filters were assessed with regard to the criteria listed above.

METHODS

Spirulina platensis and *Spirulina maxima* were grown as reported by Aiba and Ogawa (1977), either as liquid cultures or in preformed 0.5 cc polyurethane cubes by the method of Brouers et al. (1989). The impregnated cubes were removed and washed with distilled water then packed into Bio-Rad disposable chromatography columns to create an adsorptive filter prior to treatment. Filters were 5-mL bed volume, run with gravity feed and a pressure head of 15 to 20 cm. Flowrates were restricted to 0.5 mL/min. Immobilization of biomass in Ca cross-linked alginate was as described by Brouers et al. (1989).

Biomass was determined by dry weight analysis or calculated from the chlorophyll content (assuming a chlorophyll content of 1% of the dry weight). Chlorophyll was determined by the method of Mackinney (1941).

Hg was detected by the method of Snell and Snell (1948) using nitrosobenzene and potassium ferrocyanide. The complex gave an absorption at 528 nm, and was sensitive for Hg down to 10 µg/L. For lower levels of Hg, a Perkin Elmer 403 flameless atomic absorption spectrophotometer with a sensitivity down to 1 µg/L was employed. Cr was complexed with EDTA, measured at 543.5 nm and was sensitive down to 2.5 mg/L. Cu was reacted with diethyldithiocarbamate, measured at 448 nm and was sensitive down to 30 µg/L.

RESULTS

Polyurethane Foam Filters. Untreated (living algae) filters lost cellular material upon washing with various buffers and salt treatment, which resulted in clogging of the column. Moderately packed filters (where the foam was compressed to 20% of its initial volume) did not help this situation, and flowrates (gravity only) were minimal (< 5 mL/hr). Although Hg binding to the unprocessed filters was initially encouraging (1 mg/L reduced to 2 to 3 µg/L), prolonged exposure resulted in cell degradation (observed by the release of blue phycobiliproteins) and release of Hg. Filters that were heat-treated did not exhibit any of the cellular release and clogging phenomena. These filters could be either boiled for 10 min or (better) autoclaved for 20 min at 120°C. Denaturation of the cell filaments by the heat treatment probably caused them to become immobilized within the foam pores. Autoclaved (sterile) filters have been stored for up to 2 months (so far) without deleterious effects on the binding capacity. Subsequent washing removed some soluble material released by the heat processing, but loss of large cell fragments was not observed. Flowrates through the processed material were excellent (1 to 2 mL per min for a 5-mL bed volume filter) and were not a limiting factor for laboratory-scale Hg removal from aqueous solutions (10 min processing time). The quantity of immobilized biomass was estimated as 7 mg dry weight cells per 1 cc foam cube. Because of the compressibility of the foam and its excellent flow characteristics, biomass concentrations of the order of 70 mg dry weight per 1 cc of filter are presently attainable.

Binding and Stripping of Hg, Cu, and Cr. The removal of Hg by processed biomass proved to be highly modulated by the acidity of test waters. Decreasing the pH from 7 to 5 enhanced the capacity of processed algal biomass to sequester Hg by two orders of magnitude (Figure 1). The Hg binding capacity of processed *Spirulina maxima* biomass was determined to be 3 µg Hg per mg dry weight of biomass at pH 5. The filter was loaded with 1 mL 50 µM (10 mg/L) $HgCl_2$ and stripped sequentially with 10 mL aliquots of 2 M NaCl, pH 3.5, and pH 1 buffers. The column removed all detectable Hg from the test solution (residual Hg ≤ 1 µg/L). The Hg stripped from the filter by the 2 M NaCl wash had a recovery of 93%. The column has been put through three such cycles so far with no loss of binding capacity or elution performance. Low-ionic-strength (NaCl < 500 mM) or divalent cations (Ca = 10 mM) had no effect on the Hg binding. Polyurethane foam alone had no effect on Hg levels.

The polyurethane biofilter was loaded with 30 µg Cu, stripped with pH 3.5 acetate buffer, then loaded with 20 µg Cr and stripped with 2 M NaCl. The column removed both of the toxic metals from the test solution (below detectable limits). The toxic metals were stripped from the filter with recoveries of 124% and 41%, respectively. The high recovery of Cu (greater than 100% of added material) is probably due to stripping from the biomass of endogenous Cu. The low recovery of the Cr is under investigation.

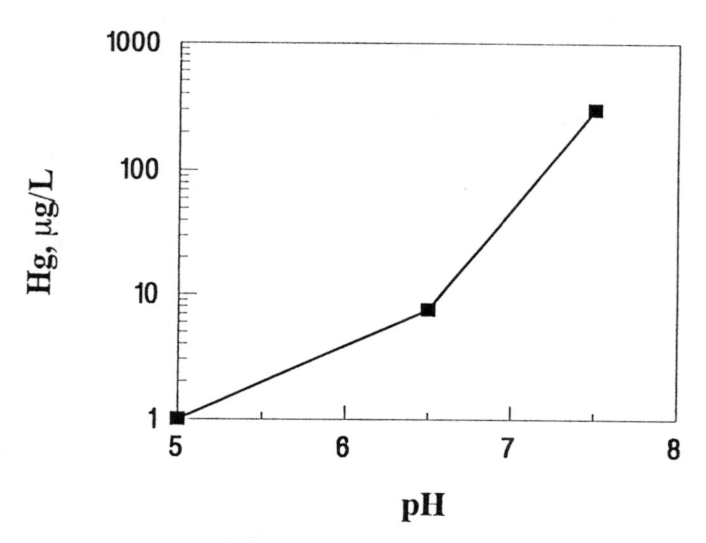

FIGURE 1. Mercury levels in synthetic test waters of various pH after passage through a processed biomass/polyurethane filter. The initial Hg level was 1 mg/L.

Alginate-Immobilized Biomass Filters. Processed *Spirulina maxima* was immobilized in alginate gel beads (0.3 cm diameter) and packed onto a column to give a 5-mL bed volume. The alginate filter was run identically to the polyurethane filters described above. The filter was loaded at pH 5 with 2 mL of 1 mg/L Hg; 85 % of the added Hg was retained by the filter, although only 5% of the theoretical Hg binding capacity was reached. This would suggest that the flowrate is more critical when using alginate beads, because there is a large void volume around the spheres and the water to be processed must permeate into the gel. Elution with various salt and acid solutions failed to strip the Hg from the filter. Experiments with alginate alone suggested that the Hg was exchanging with the divalent cation cross-linkers in the alginate matrix rather than binding to the processed biomass and could not be displaced by moderate salt treatments. Prolonged incubation of the alginate in water with low divalent cation content (< 100 µM) caused destabilization of the alginate beads, with release of the immobilized biomass and Hg. Moreover, anions that competed with the alginate for the cross-linking divalent cations (such as phosphate) rapidly destabilized the alginate matrix.

CONCLUSION

Of the two types of immobilizing matrices examined, the preformed polyurethane foam outperformed the Ca cross-linked alginate gels in most respects, particularly with respect to stability and recycling performance. From the binding capacity data obtained with the polyurethane foam filters (70 mg biomass per 1 cc of filter and 3 µg Hg bound per mg biomass), we can calculate the binding capacity of a laboratory-scale adsorptive filter. A 1-L filter would have the capacity to bind 210,000 µg Hg, which is equivalent to processing 21,000 L of water contaminated with 10 µg/L Hg down to a level below 1 µg/L. The time frame for such removal is 36 days, calculated from the present bench-top experiments and using a 1-L biofilter. Obviously, increasing the biofilter volume would decrease the processing time.

ACKNOWLEDGMENTS

This work was supported by the Office of Technology Development, U.S. Department of Energy, and by the Director, Office of Energy Research, Division of University and Science Education Programs of the U.S. Department of Energy under Contract DE-AC-76SF00098.

REFERENCES

Aiba, S., and T. Ogawa. 1977. "Assessment of growth yield of a blue-green alga, *Spirulina platensis*, in axenic and continuous culture." *J. Gen. Microbiol. 102*: 179-182.

Brouers, M., D. J. Shi, and D. O. Hall. 1989. "Immobilization methods for *Cyanobacteria* in solid matrices." *Methods in Enzymol. 167*: 629-638.

Mackinney, G. 1941. "Adsorption of light by chlorophyll solutions." *J. Biol. Chem.* 140: 313-322.

Snell, F. D., and E. C. Snell. 1948. In *Colorimetric Methods of Analysis*, Vol. IIA. Van Nostrand, New York, NY.

MICROBIAL MAT DEGRADATION OF CHLORDANE

J. Bender, R. Murray, and P. Phillips

INTRODUCTION

Despite a two-decade ban on chlordane, this compound persists in soils and in lake and river sediment. A 1986 review by the U.S. Environmental Protection Agency (EPA) concluded that chlordane and heptachlor are likely human carcinogens and tumor promoters (EPA 1986a). Chlordane is easily absorbed through the skin, is known to pass through the placenta, and can be found in breast milk (Stacey & Tatum 1985). Further, extensive exposure to chlordane may result in chlordane poisoning leading to various nervous system disorders. From the ecological perspective, the persistence of chlordane in the environment is well known, but its long-range impact on soil bacteria, flora, and fauna has not been defined.

Because of the toxic nature of chlordane and similar compounds, the present study was designed as a model for the degradation of cyclic chlorinated hydrocarbons. In this work, a microbial mat consortium dominated by cyanobacteria or blue-green algae (*Oscillatoria* sp.) was developed and used in chlordane degradation experiments.

METHODS

Microbial mats were constructed specifically for chlordane degradation by first acclimating the cyanobacterium, *Oscillatoria* sp., and the associated chemoautotroph, *Chromatium* sp., to elevated concentrations (> 2,000 mg/L) of chlordane. All microorganisms used in this research were wild strains originally isolated over a two-year period from various uncontaminated soil and water samples. Chlordane-tolerant strains were used to inoculate for the development of multispecies mats. To construct a mat, ensiled grass clippings (7 g/L wet weight) were added to enriched medium (Allen & Arnon 1955) together with the chlordane-tolerant microbes described above. Within 7 to 10 days multispecies, chlordane-tolerant mats were formed. These mats, containing autotrophs and heterotrophs, were tightly annealed in a gel matrix that floated on the surface of the water column. Sections of the mats were added to covered beakers containing a 50-mL water column.

After mats began to show growth in the beakers, 200 mg/L chlordane (without solvent) was added to the water column. Although chlordane is not soluble in water, the active sequestering of mat filaments drew the droplets of chlordane into the microbial matrix, thereby avoiding the problem of achieving chlordane solubility for cell contact.

Mat-chlordane beakers were prepared to provide material for a daily sampling over a 5-day experimental period. On each sampling day, all materials from the beakers (mat and water column) were harvested and extracted. Solid portions (mats) were extracted with Soxhlet extraction (EPA 1986b) using six reflux cycles (J. Gould, personal comm., 1992), and water columns were extracted in chloroform using a separatory funnel. Combined extracts from mats and water columns were concentrated on a rotary evaporator to a final volume of 1 mL and filtered to remove sediments. Concentrates were analyzed for chlordane concentrations with high-performance liquid chromatography (HPLC; Beckman Instruments with System Gold programmable detector equipped with a 10-cm Whatman RCA II column). The mobile phase was a methanol:water 60:40 gradient with a total retention time of 20 min.

In separate experiments, various microbial components of the mats were assessed for chlordane-degrading capacity. Controls contained heat-sterilized growth medium with ensiled grass.

RESULTS AND DISCUSSION

Reduced peaks of the chlordane chromatograms indicated rapid reduction in chlordane concentration (Figure 1). This degradation was apparent in the first day of mat exposure and continued at a constant rate through day 6. The degradation rate of chlordane (200 mg/L) is given in Figure 2.

After 6 days the microbial diminished in the water column and sedimented at the bottom of the beaker. At the same time, complete microbial mats remained intact and showed active growth for extended periods (>30 days). Although nutrient supplements were required to maintain the bacteria outside of the mat system, no supplements were necessary for the maintenance of the mat. The self-maintenance of the mat, resulting from the capacity to fix both nitrogen and carbon, may be a distinct advantage for low-cost bioremediation.

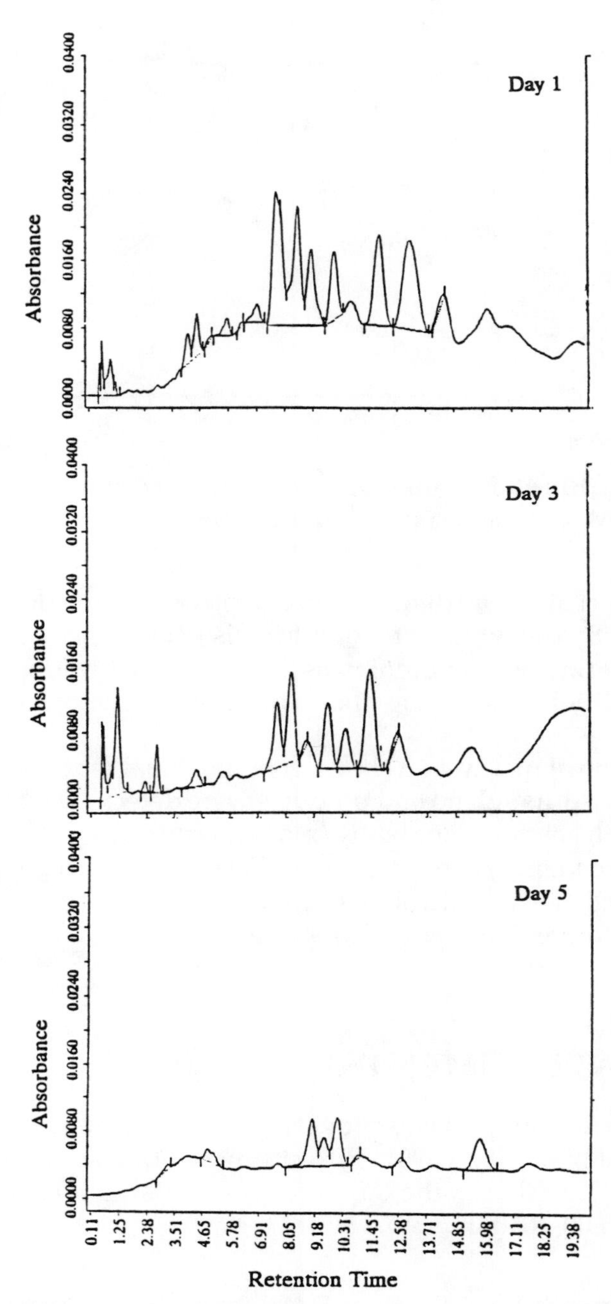

FIGURE 1. HPLC chromatogram series from a 5-day mat treatment of chlordane. Chlordane (200 mg/L) was treated with microbial mats. Entire treatment systems (mats and water columns) were harvested, extracted, and analyzed on days 1, 3, and 5.

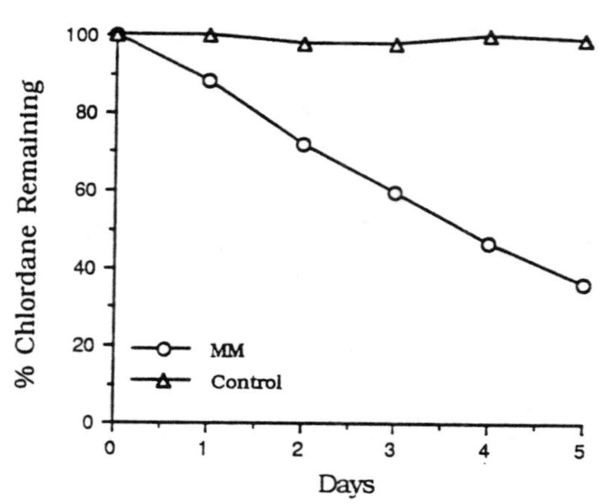

FIGURE 2. Chlordane removal from the water column by microbial mat. MM = constructed microbial mat.

Kennedy et al. (1990) demonstrated a relatively slow degradation rate of chlordane by the white-rot wood fungus *Phanerochaete chrysosporium*. In these experiments the organisms were cultured under nutritionally limited conditions obligating the production of chlordane-degrading lignase.

The microbial mat system shows good potential for rapid degradation of chlordane. Although no end products were detected by HPLC, further experiments in gas chromtography/mass spectroscopy (GC/MS) analyses are planned to identify these products. Current research is (1) identifying the specific consortial microbes responsible for chlordane degradation and (2) developing biological assays to test the toxicity of metabolic end products.

ACKNOWLEDGMENTS

Infrastructure support was provided by U.S. Environmental Protection Agency Grant No. CR81868901. Project support for chlordane degradation analysis was provided by the U.S. Agency for International Development DAN-5053-G-00-1050-00.

REFERENCES

Allen, M. B., and D. I. Arnon. 1955. "Studies on Nitrogen-Fixing Blue-Green Algae 1. Growth and Nitrogen Fixation by *Anabaena cylindrica*." *Lemm. Pl. Physiol. 30*: 366-372.

Kennedy, W. D., D. S. Aust, and J. A. Bumpus. 1990. "Comparative Biodegradation of Alkyl Halide by the White Rot Fungus *Phanerochaete chrysosporium* (BKM-F-1767)." *Applied and Environmental Microbiology 56(8)*: 2347-2353.

Stacey, C. I., and T. Tatum. 1985. "House Treatment with Organochlorine Pesticides and Their Levels in Human Milk." *Bulletin of Environmental Contamination and Toxicology 35*: 202-209.

U.S. Environmental Protection Agency. 1986a. "Carcinogenicity Assessment of Chlordane and Heptachlor/Heptachlor Epoxide." *Office of Health and Environmental Assessment*, Washington, DC.

U.S. Environmental Protection Agency. 1986b. "Test Method for Evaluating Solid Waste." Volume 1B: Laboratory Manual 846, Physical/Chemical Methods, Washington, DC.

AUTHOR LIST

B. C. Alleman
Battelle
505 King Avenue
Columbus, OH 43201-2693 USA

T. Barkay
U.S. Environmental Protection Agency
Environmental Research Laboratory
One Sabine Island Drive
Gulf Breeze, FL 32561 USA

L. J. Barnes
Shell Research, Ltd.
Sittingbourne Research Centre
Sittingbourne, Kent ME9 8AG
England UK

J. Bender
Research Center for Science and Technology
Clark Atlanta University
Box 296
James P. Brawley Drive at Fair Street, S.W.
Atlanta, GA 30314 USA

J. L. Bolis
Environmental Sciences and Engineering Ecology Department
Colorado School of Mines
Golden, CO 80401 USA

B. B. Buchanan
Department of Plant Biology
University of California
Berkeley, CA 94720 USA

C.J.N. Buisman
Paques Environmental Technology
Box 52, 8560 AB Balk
THE NETHERLANDS

K. Chellman
Syntex, Inc.
3401 Hillview Avenue
P.O. Box 10050
Palo Alto, CA 94303 USA

L.C.M. Commandeur
Department of Environmental and Toxicology Chemistry
University of Amsterdam
Amsterdam
THE NETHERLANDS

L. Copley-Graves
Battelle
505 King Avenue
Columbus, OH 43201-2693 USA

L. Diels
Lab. of Genetics and Biotechnology
Vlaamse, Instelling voor Technologisch Onderzoek
Boeretang 200
2400, Mol
BELGIUM

U. Ekanemesang
Clark Atlanta University
James P. Brawley Drive at Fair Street, S.W.
Atlanta, GA 30314 USA

W. T. Frankenberger, Jr.
Soil & Environmental Sciences
University of California
Riverside, CA 92521 USA

I. V. Fry
Lawrence Berkeley Laboratory
Mail Stop 70-193A
Berkeley, CA 94720 USA

C. F. Gökcay
Middle East Technical University
Department of Environmental Engineering
06531, Ankara TURKEY

B. Graves
Clark Atlanta University
Research Center for Science and Technology
James P. Brawley Dr. at Fair St., S.W.
Atlanta, GA 30314 USA

K. Y. Henry
Levine-Fricke, Inc.
1900 Powell Street, 12th Floor
Emeryville, CA 94608 USA

J. L. Kipps
California Department of Water Resources, San Joaquin District
3374 East Shields Avenue
Fresno, CA 93726 USA

T. Leighton
Biochemistry and Molecular Biology
University of California
Berkeley, CA 94720 USA

G. M. Leong
Levine-Fricke, Inc.
220 South King Street, Suite 1290
Honolulu, HI 96813 USA

D. S. Lipton
Levine-Fricke, Inc.
1900 Powell Street, 12th Floor
Emeryville, CA 94608 USA

D. A. Martens
Soil & Environmental Sciences
University of California
Riverside, CA 92521 USA

R. J. Mehlhorn
Heavy Metal and Free Radical Toxicology Group
Lawrence Berkeley Laboratory
Mailstop 70 - 193A
Berkeley, CA 94720 USA

M. Mergeay
Lab. of Genetics and Biotechnology
Vlaamse, Instelling voor Technologisch Onderzoek
Boeretang 200
2400, Mol
BELGIUM

R. E. Moon
Geraghty and Miller, Inc.
14497 N. Dale Mabry Hwy., Suite 115
Tampa, FL 33629 USA

R. Murray
Clark Atlanta University
Research Center for Science and Technology
Box 271
Atlanta, GA 30314 USA

E. Nyer
Geraghty & Miller, Inc.
14497 N. Dale Mabry Hwy., Suite 115
Tampa, FL 33629 USA

S. Önerci
Etibank General Directorate, Sihhiye
Ankara
TURKEY

J. R. Parsons
Department of Environmental and Toxicology Chemistry
University of Amsterdam
Amsterdam
THE NETHERLANDS

P. Phillips
Research Center for Science and Technology
Clark Atlanta University
James P. Brawley Drive at Fair Street, S.W.
Atlanta, GA 30314 USA

J. S. Reynolds
Environmental Sciences and Engineering Ecology Department
Colorado School of Mines
Golden, CO 80401 USA

S. Rodriguez-Eaton
Clark Atlanta University
Research Center for Science and Technology
James P. Brawley Dr. at Fair St., S.W.
Atlanta, GA 30314 USA

A. Ryngaert
Lab. of Genetics and Biotechnology
Vlaamse, Instelling voor Technologisch Onderzoek
Boeretang 200
2400, Mol BELGIUM

E. Saouter
University of West Florida
Pensacola, FL 32514 USA

P.J.M. Scheeren
Budelco BV
Hoofdstraat 1, NL 6024 A A
Budel - Dorplein
THE NETHERLANDS

L. A. Smith
Battelle
505 King Avenue
Columbus, OH 43201-2693 USA

D. Springael
Lab. of Genetics and Biotechnology
Vlaamse, Instelling voor Technologisch Onderzoek
Boeretang 200
2400, Mol BELGIUM

J. M. Thomas
Levine-Fricke, Inc.
1900 Powell Street, 12th Floor
Emeryville, CA 94608 USA

R. Turner
Environmental Science Division
Oak Ridge National Laboratory
Oak Ridge, TN 37831 USA

D. M. Updegraff
Chemistry & Geochemistry
Colorado School of Mines
Golden, CO 80401 USA

J. van Thor
Department of Environmental and Toxicology Chemistry
University of Amsterdam
Amsterdam
THE NETHERLANDS

Y. Vatcharapijarn
Clark Atlanta University
Research Center for Science and Technology
James P. Brawley Drive at Fair Street, S.W.
Atlanta, GA 30314 USA

T. R. Wildeman
Chemistry & Geochemistry
Colorado School of Mines
Golden, CO 80401 USA

INDEX